Research in Parapsychology 1992

Abstracts and Papers from the Thirty-Fifth Annual Convention of the Parapsychological Association, 1992

Emily Williams Cook
Editor

The Scarecrow Press, Inc.
Lanham, Md., & London

SCARECROW PRESS, INC.

Published in the United States of America
by Scarecrow Press, Inc.
4720 Boston Way
Lanham, Maryland 20706

4 Pleydell Gardens, Folkestone
Kent CT20 2DN, England

British Cataloguing-in-Publication Information Available

Library of Congress Cataloging-in-Publication Data

Parapsychological Association. Convention (35th : 1992 : Las Vegas, Nev.)
Research in parapsychology 1992 : abstracts and papers from the Thirty-Fifth Annual Convention of the Parapsychological Association, 1992 / [edited] by Emily Williams Cook.
p. cm.
Papers originally presented at the 35th Annual Convention of the Parapsychological Association, Las Vegas, Nev., Aug. 9–13, 1992.
Includes bibliographical references and index.
1. Parapsychology—Congresses. I. Cook, Emily Williams. II. Title.
BF1021.P28 1992 133—dc20 95-35064 CIP

ISBN 0–8108–3041–8 (cloth : alk. paper)

Printed in the United States of America

∞ ™ The paper used in this publication meets the minimum requirements of American National Standard for Information Sciences—Permanence of Paper for Printed Library Materials, ANSI Z39.48–1984.

Contents

3. Panel Discussion

4. Poster Session

5. Invited Addresses

6. Presidential Address

Preface: The Thirty-Fifth Annual Convention

The 35th Annual Convention of the Parapsychological Association was held August 9–13, 1992, at the Hacienda Hotel, Las Vegas, Nevada. About 101 people attended.

The program consisted of fourteen papers, three posters, one symposium, one panel discussion, three invited addresses (by John Beloff, Daryl Bem, and William Braud), and the Presidential Address by John Palmer. An abbreviated or (in the case only of the three addresses printed here) full version of each presentation is included in this volume. In addition, the program included the J. B. Rhine Lecture following the banquet (given this year by Jeanne Achterberg and entitled "Imagery and Medicine: Research and Applications"); a tribute to Hella Hammid, who died in 1992; and three informal workshops ("A Demonstration of Topographic Brain Mapping," given by Norman S. Don, Bruce E. McDonough, and Charles A. Warren; "Parapsychology in Russia: Past, Present, and Future," given by Eduard K. Naumov; and "On-Line Archive of Psi-Data," given by Dick J. Bierman).

The members of the Program Committee were Emily Williams Cook, Chair, and Carlos S. Alvarado, Gary Heseltine, Ephraim I. Schechter, and Gertrude R. Schmeidler. Stephen Braude, James McClenon, and Marilyn Schlitz provided helpful assistance as specialist consultants. Laura F. Knipe, Arrangements Chair, did her usual excellent job of locating the conference site and making all associated arrangements.

Emily Williams Cook
Program Chair
Division of Personality Studies
University of Virginia
Box 152 Health Sciences Center
Charlottesville, VA 22908

Editorial Introduction

In this volume of *Research in Parapsychology,* the fourteen papers are arranged by topic and not in the order in which they were presented at the convention. Because of this arrangement, the names of the persons chairing each session cannot be listed with the papers, and so I thank here the following people who served as chairpersons: Carlos S. Alvarado, John Beloff, Stephen Braude, Deborah Delanoy, Charles Honorton, Stanley Krippner, Michaeleen Maher, Robert McConnell, Robert Morris, K. Ramakrishna Rao, Marilyn Schlitz, Gertrude Schmeidler, and Nancy Zingrone.

At this year's convention, audience discussion of each paper was stimulated by comments from a discussant, and I would like to thank the following people who played this role: John Beloff, Daryl Bem, Rick Berger, Kathy Dalton, Deborah Delanoy, Stanley Krippner, Edwin May, Robert Morris, Marilyn Schlitz, and Gertrude Schmeidler.

The Invited Addresses and the Presidential Address have been printed in full. All other contributions, however, are significantly reduced versions of the full conference papers. In many cases, length restrictions have required authors to eliminate over half of their original reports. Therefore, it should not be assumed that the articles contained herein are complete descriptions of the research being reported. The mailing address of one author of each abstract is included, when possible, so that readers can contact an author for a copy of the complete report.

Readers should also be aware that although the Convention Program Committee does evaluate all submissions to determine their acceptability for presentation, this review process must be conducted within a limited time period. Time constraints are such that papers cannot receive the level of evaluation given papers submitted to journals.

Emily Williams Cook
Editor

Abbreviations

The following abbreviations are used in reference citations:

EJP	*European Journal of Parapsychology*
IJP	*International Journal of Parapsychology*
JASPR	*Journal of the American Society for Psychical Research*
JP	*Journal of Parapsychology*
JSPR	*Journal of the Society for Psychical Research*
PASPR	*Proceedings of the American Society for Psychical Research*
PSPR	*Proceedings of the Society for Psychical Research*
RIP 19xx	*Research in Parapsychology (for the year indicated)*
ZP	*Zeitschrift für Parapsychologie*

Erratum

With respect to the abstract entitled *A Ganzfeld Experiment with Subliminal "Sending"* (*RIP 1990*, pp. 50-55), we have determined that the statistical significance of the overall psi-missing effect on the ESP task, expressed therein as "$p = .03$, two-tailed", is more properly expresses as "$p = .94$, one-tailed". The latter corresponds to the figures reported in the *Journal of Parapsychology*, 1993, 57, 241-257.

John Palmer
H. Kanthamani

Part 1: Papers

PHYSIOLOGICAL STUDIES

Further Studies of Autonomic Detection of Remote Staring: Replications, New Control Procedures, and Personality Correlates

William Braud†, Donna Shafer, and Sperry Andrews (Mind Science Foundation, 8301 Broadway, Suite 100, San Antonio, Texas 78209)

In a paper presented at the 1990 Parapsychological Association Convention (Proceedings of Presented Papers, 1990, 14–28), we reviewed experimental attempts to assess persons' accuracy in consciously detecting when they were being stared at by someone beyond the range of their conventional senses. We also reported results of our own new experiments in which a more unconscious autonomic nervous system reaction was used to assess accuracy of detection of staring (remote attention). We reasoned that autonomic reactions might be less distorted by higher cognitive processes and therefore might provide a purer and more sensitive indicator. In two experiments, one person directed full attention to another (distant) person's image on the monitor of a closed circuit television system (used to eliminate the possibility of subtle sensory cues), while the latter person's spontaneous electrodermal

†The dagger here and in subsequent papers denotes the speaker(s).

activity (a sensitive index of sympathetic nervous system activity) was monitored objectively by a computer system during randomly interspersed staring and nonstaring periods; the staree was blind regarding the number, timing, and sequencing of the two types of periods. We found evidence for significant autonomic discrimination of the two types of periods. As judged by effect sizes, unconscious autonomic detection did indeed appear to yield stronger effects than did previous conscious verbal or motoric detection assessments.

Here, we replicate and extend our previous findings, with two improvements. In Replication 2, we added an empirical control condition (a "sham control") in which we treated sessions and data as we did for real staring sessions, but staring did not, in fact, occur; this provided a baseline assessment of the likelihood of obtaining chance discriminations of otherwise equivalent session segments. The second improvement was the introduction of an additional personality assessment, a Social Avoidance and Distress (SAD) scale.

METHOD

Subjects

Thirty volunteer participants (22 females and 8 males) served as "starees" for Replication 1, and 16 volunteers (5 males and 11 females) participated as starees for Replication 2. For Replication 1, each starer worked with 10 starees, and results for all 30 starees were pooled for analysis. In Replication 2, one starer worked with all 16 starees. The starers of Replication 1 were three undergraduate psychology students (2 females and 1 male) from a local college. D. S. served as starer for Replication 2.

Apparatus

The experimental apparatus was identical to that described in our 1990 report and consisted of silver/silver chloride palmar electrodes, a skin-resistance amplifier, and an analog-to-digital converter interfaced with a microcomputer. A color video camera in the staree's room permitted the staree to be viewed by the starer in a distant room without sensory cueing.

Procedure

The starer and staree were stationed in separate rooms in entirely different suite areas; this precluded conventional sensorimotor communication between starer and staree under the conditions of the experiment. The camera was active continuously throughout the session.

Forty-six sealed envelopes had been previously prepared by W. B., who had used a computer's random algorithm to generate the random sequence of the 10 staring and 10 nonstaring periods for each session. If the random sequence indicated a staring period, the experimenter/starer silently swiveled his or her chair around so that it faced the television monitor and stared intently at the staree's monitor image throughout the 30-second recording periods. During nonstaring periods, the experimenter/starer kept the chair turned away from the monitor, so that the monitor's screen could not be seen.

Throughout the session, the experimenter/starer was provided with no information about the staree's ongoing electrodermal activity; the latter was continuously and automatically assessed by the computer system. The equipment sampled the staree's rectified spontaneous phasic skin resistance responses 10 times each second for the 30 seconds of a recording epoch and averaged these measures. At the end of the experimental session, the computer printed the electrodermal results for each of the 20 recording periods.

For Replication 2, instead of 20 recording epochs, a session consisted of 32 recording epochs. One half of a session was an "experimental" half and included 8 staring and 8 nonstaring periods (randomly interspersed). The other half of a session served as a new empirical "sham control" and included eight sham or pseudostaring periods and eight nonstaring periods (also randomly interspersed). The experimental half provided a comparison of true staring versus the absence of staring. The sham control half provided data that were treated in the same manner as the experimental data, but actual staring did not take place during the sham/pseudostaring periods. The staree had no way of knowing, conventionally, when the "real" periods and when the sham periods were in effect and was monitored in an identical fashion throughout both types of periods.

Personality Assessments

For Replication 1, we administered the Social Avoidance and Distress (SAD) scale, which measures social-evaluative anxiety (the experience of distress, discomfort, fear, and anxiety in social situations) and deliberate avoidance of social situations. For Replication 2, the SAD scale was used along with the Myers-Briggs Type Indicator (MBTI, Form F). For this study, we were especially interested in the MBTI extraversion-introversion scale because of its possible relationships with SAD scoring and with remote staring detection effects in this psi-mediated social (staring) context.

RESULTS

Primary Analyses

For each of the 30 sessions for Replication 1, a total score was calculated for all 20 recording periods (10 staring and 10 nonstaring). This total score was divided into the sum of the electrodermal activity scores for the 10 staring (S) periods; the process was repeated for the 10 nonstaring (N) periods. In the absence of a remote staring effect, these two ratios [S/(S+N), N/(S+N)] should approximate 50%. A remote staring effect would be indicated by a significant departure of the scores from the 50% mean chance expectation (MCE). A single mean t test was used to assess the departure of the ratios from MCE (50%). A similar analysis was performed for the 16 sessions of Replication 2, with 8 staring and 8 nonstaring periods contributing to each session score.

For Replication 1, mean electrodermal activity percentages were 45.15% (for the staring periods) and 54.85% (for the nonstaring periods), rather than the 50%/50% expected by chance. This scoring rate yielded a single mean $t = 1.85$, 29 df, $p = .06$, two-tailed, and an effect size $(r) = .34$. For Replication 2, mean electrodermal activity percentages were 45.66% (for the staring periods) and 54.34% (for the nonstaring periods), rather than the 50%/50% expected by chance. This scoring rate yielded a single mean t = 2.08, 15 df, p = .05, two-tailed, and an effect size (r) = .47. Thus, for both replications, the autonomic discrimination took the form of reduced spontaneous electrodermal activity during staring periods, compared with nonstaring periods. For the 16 sham control session scores of Replication 2 (each based on 8 pseudostaring and 8 nonstaring periods) the mean electrodermal activity percentages were, as expected, virtually identical to the 50%/50% values expected on the basis of chance. Here the scoring rates were 49.16% (for the pseudostaring periods) and 50.84% (for the nonstaring periods). This scoring rate yielded a single mean t = 0.30, 15 df, p = .76, two-tailed, and an effect size (r) = .08.

Secondary Analyses

Linear correlation coefficients (Pearson r's) were calculated in order to determine the interrelationships among the magnitude of the remote staring detection effect, SAD scoring, and MBTI extraversion-introversion (E/I) scoring. To study the remote staring detection-SAD relationship, Pearson r's were computed for the percent electrodermal activity occurring during the staring periods versus the SAD scores (expressed as a percentage of the highest possible SAD score) for Replication 1, for Replication 2, and for the Sham Control sessions.

For Replication 1, the magnitude of the remote staring detection effect (the degree of "calming" during the staring periods) was significantly and positively correlated with degree of social avoidance and distress (r = .36, 28 df, p = .05, two-tailed); a similar trend was found for Replication 2 (r = .43, 14 df, p = .09, two-tailed). For the Sham Control sessions, on the other hand, this same correlation was small, negative, and nonsignificant (r = −.12, 14 df, p = .66, two-tailed).

To study the remote staring detection-introversion relationship, Pearson r's were computed for the percent electrodermal activity during the staring periods versus the continuous score for the MBTI introversion scale, for Replication 2, and for the Sham Control sessions (the MBTI was not administered for Replication 1).

For Replication 2, there was a strong, positive, and highly significant correlation between the magnitude of the remote staring detection effect and the staree's degree of MBTI introversion ($r = .68$, 14 df, $p = .0037$, two-tailed). No such correlation occurred for the Sham Control segment of the experiment ($r = .16$, 14 df, $p = .55$, two-tailed).

DISCUSSION

Evidence for autonomic discrimination of remote watching or staring was obtained in all four studies conducted thus far. This evidence reached statistical significance in the two 1990 studies and in Replication 2 of the present paper and very closely approached significance ($p = .06$, two-tailed) in Replication 1 of the present paper. The effect sizes were all relatively large, ranging from .34 to .57. The absence of a similar effect in the special Sham Control trials provides another indication that the effect obtained in the "real" trials was not artifactual. The significant correlations that obtained between the remote staring detection effect and the two personality variables of SAD and MBTI introversion in the real experiments but not in the Sham Control segment provide additional evidence for the reality of the remote staring detection effect and also relate the magnitude of the effect to certain psychological variables.

Topographic Brain Mapping of a Selected Subject Using the ESPerciser™: A Methodological Demonstration and an Exploratory Analysis

Norman S. Don†, Bruce E. McDonough, and Charles A. Warren (Kairos Foundation and the University of Illinois at Chicago). Address correspondence to Norman S. Don, Ph.D., Kairos Foundation, 405 N. Wabash Ave., Suite 2201, Chicago, IL 60611.

INTRODUCTION

Topographic brain mapping is a new method of representing EEG data from multiple recording sites. Topographic maps indicate, by means of colored contour graphs of electrical power, the functional state of the brain. That is, during a task, the maps display which areas of the brain are activated, the frequencies present, and the power at these frequencies. Topographic mapping is becoming increasingly important for research and clinical applications. Interested readers should see Duffy, Iyer, and Surwillo (*Clinical Electroencephalography and Topographic Brain Mapping: Technology and Practice,* Springer-Verlag, 1989).

Presently, a selected subject, Olof Jonsson, was tested on a computerized guessing task, the ESPerciser™, developed at the Psychophysical Research Laboratories. This forced-choice guessing task presents graphical images on a video monitor. Targets and decoys are randomly selected by a hardware random number generator, and the recording of the subject's responses is completely automated. Jonsson is a controversial psychic who has previously performed well on the ESPerciser™ (See Don, McDonough, & Warren, *JP,* 1992, 87–96).

METHOD

Task

The ESPerciser™ software was modified so as to present four graphic stimuli in the center of the video screen in serial order. The stimulus set was limited to four of the five Zener symbols: star, wave, cross, and circle. The modified program was run on an Apple IIGs computer using an internal hardware random number generator (RNG). Randomness tests conducted within two days prior to and after the present experiment and similar analyses of RNG output performed routinely at the laboratory over many months of periodic testing did not indicate consistent deviations from theoretical expectations.

Testing Procedure

The experiment was conducted in four sessions held on four consecutive days. The number of runs per day was left flexible because of the subject's health status. One run of twenty-four trials was completed the first day, five runs the second, and three runs on days three and four.

While the subject was using the ESPerciser™, his EEG was recorded using a Lexicor Medical Technology, Inc. NeuroSearch-24 topographic brain mapper. This equipment consisted of nineteen AC-coupled amplifiers for EEG recording (time constant approximately 0.3 secs). The sampling rate was 256/sec. Electrodes were applied using an electrode cap made by Electro-Cap International, Inc.

Data Analysis

EEG data were visually inspected for eye movement and other artifacts, blind to the subject's performance. Epochs containing artifacts were excluded from further analyses. For data analyses, four one-second EEG epochs preceding the subject's target selection on the ESPerciser™ were each spectrally analyzed and their spectra averaged. For each channel, power in the theta (4–7 Hz), alpha (8–13Hz), and beta (14–55 Hz) frequency bands was computed. Power was expressed as microvolts squared.

RESULTS

Behavioral Performance

There were 68 hits out of 288 trials ($p = .727$, one-tailed exact binomial). Post-hoc analyses revealed that in Run 1 the subject scored 12/24 hits ($p < .007$). We refer to this as the "psi-hitting run." For Run 5 he scored 2/24 hits ($p < .040$), which we called the "psi-missing run." Hit rates for the remaining 10 runs were close to MCE. The three runs (4, 7, and 10) with hit rates exactly equal to MCE were defined as "chance runs."

Topographic Maps

The topographic maps produced by the Lexicor system were 12-color contour maps derived from a linear interpolation algorithm employing power values at four nearest neighbors. For each run, one map for each of three EEG bandpasses was displayed.

For Run 1 (possible psi-hitting) there is a clear left-to-right power gradient in both the theta and alpha bands and, to a large degree, for the beta band. A pattern of maximum or near maximum powers is found at the right lateral portions of the scalp. For theta, this includes F8 and T4; for alpha, there is a projection of this area toward Pz; for beta, there is a strong anterior-posterior sector which includes F8, T4, T6 and extends toward the midline. All minima were found at the left lateral portions, centered at the left anterior temporal site (T3). The general picture for all three bands is of right hemisphere power maxima with a fairly uniform decrease of power along an approximately right-to-left transverse axis.

For Run 5 (possible psi-missing), the theta band shows a reversal of the power pattern in Run 1, with the region of maximum power covering a roughly triangular sector of the scalp. This region covers the left lateral frontal lobe in a wedge extending from the vertex (Cz) and opening out to include the left lateral scalp (T3-F7, F7, F7-Fp1). This wedge is surrounded by high power areas which decrease in power with increasing distance from the wedge, i.e., toward the right. The power minimum is found at the right temporal locus (T4). The alpha band shows minimum power at the right temporal site (T4) with increasing power from right to left. Power maxima are at the left posterior temporal (T5) and in an area including the central and parietal midline (Cz and Pz). In the T3 through F7 loci near maximum power is also found. The beta band shows maximum power at the left posterior periphery: O1 and T5. Although there is a high power region at T4, it is much smaller than in Run 1. A region of minimum power extends anterior-posterior and laterally right and left covering over two-thirds of the scalp. The general picture is of very large areas of the brain at low power. Also, the left posterior power maximum and the lower power secondary maximum at T4 is a reversal of the pattern in Run 1.

In general, maps of the three chance level runs (Runs 4, 7, and 10) did not exhibit strong, consistent right-left or left-right power gradients.

Statistical Analysis of EEG Power

We performed a set of t tests on power in the theta, alpha, and beta frequency bands for selected electrode sites in order to statistically assess the apparent differences observed in the topographic EEG maps. Due to the large number of electrodes used, we examined only the right temporal (T4) and left temporal (T3) sites (selected before seeing the data on the basis of previous findings from this laboratory indicating the involvement of EEG activity at these sites in psi performance). The initial comparisons were made between Run 1 (psi-hitting) and Run 4 (the first occurring run with hits equalling MCE). At T4, Run 1 had greater power than Run 4 for the theta ($t = 4.17$, 46 df, $p \leq .001$), alpha ($t = 6.02$, 46 df, $p \leq .001$), and beta ($t = 4.16$, 46 df, $p \leq .001$) frequency bands. At T3, Run 1 had less power than Run 4 for theta ($t = -5.95$, 46 df, $p \leq .001$), alpha ($t = -6.19$, 46 df, $p \leq .001$), and beta ($t = -4.08$, 46 df, $p \leq .001$). Confirmatory analyses (not reported here) performed between Run 1 and Run 7 and between Run 1 and Run 10 yielded comparable results. Runs 7 and 10 were also chance performance runs, as defined above.

A second set of t tests between Run 5 (psi-missing) and the three chance runs yielded a generally opposite pattern of results, although weaker. Theta power at T4 was marginally less for Run 5 than for Run 4 ($t = -1.69$, 46 df, $p \leq .097$) and significantly less for alpha ($t = -2.28$, 46 df, $p \leq .027$) but greater in Run 5 than in

Run 4 for beta ($t = 2.69$, 46 df, $p \leq .010$). Confirmatory analyses (not reported here) comparing power at T4 in Run 5 versus Runs 7 and 10 yielded similar results. At T3, power in the theta, alpha, and beta bands did not significantly differ (all $p > .05$) between Run 5 and Run 4. However, comparisons between Run 5 and Runs 7 and 10 (also not reported here) indicated that Run 5 may have had more power at T3 than these later runs in all three frequency bands.

DISCUSSION

As far as we know, this is the first published psi study employing frequency-domain topographic mapping. These maps qualitatively suggest that psi-hitting, psi-missing, and chance-level performance are closely related to topographic patterns of power at certain frequencies recorded at the scalp. The statistical analyses of EEG power, although limited to two recording sites, supported the apparent differences observed in the topographic maps. Because these patterns were present for all trials (hit and miss alike) within each run, it is possible that the differences between runs were due to functional states of the brain which were conducive to psi-hitting, psi-missing, or chance performance. This interpretation is offered tentatively, however, because the performance data did not provide unambiguous evidence of psi. Also, because the present comparisons were made between runs, rather than between hit versus miss trials within runs, the observed differences in EEG patterns between runs may have been due to spurious differences in psychological state such as mood, alertness, fatigue, etc.

In summary, these maps, based on limited data, qualitatively suggest that psi-hitting, psi-missing, and chance behavior may be closely related to topographic patterns of power at certain frequencies recorded at the scalp. However, further research is needed to confirm these findings. In any event, it seems clear that topographic brain maps, which incorporate EEG data from numerous recording sites in an integrated and readily understood manner, may be useful in psi research.

Mind, Brain, and Behavior: A Preliminary Study of Their Interrelationships During Performance of a Computerized Psi Task

Bruce E. McDonough†, Norman S. Don, and Charles A. Warren (Kairos Foundation and University of Illinois at Chicago). Address correspondence to Bruce E. McDonough, Ph.D., Kairos Foundation, 405 N. Wabash Avenue, Suite 2201, Chicago, IL 60611.

BACKGROUND

Over a period of more than two years, a series of seven psi experiments was conducted in which one of the present investigators (BEM) served as subject. Each study comprised 100 trials on the ESPerciser™, a forced-choice psi task developed at the Psychophysical Research Laboratories. Each trial began with an "impression period" of subject-determined length, followed by the presentation of four visual stimuli on the (Apple II) computer's video screen. After the subject's response, the target for that trial, randomly selected by RNG, was presented as feedback. The stimulus pool comprised a wide variety of graphical, cartoon-like figures.

In Study 1, the subject scored 36 hits out of 100 trials ($p = .009$, one-tailed exact binomial). Unfortunately, no electroencephalographic (EEG) recordings were made in Study 1. Encouraged by these results, the authors conducted Study 2 to study the EEG correlates of the psi performance that we expected to observe again. This time, however, the subject scored only 29 hits ($p = .206$). Because performance appeared to be at chance level, the EEG data were not (at that time) analyzed. Moreover, because EEG studies require greatly increased time and effort relative to studies of performance only, subsequent studies were planned without EEG recording until such time as successful psi performance appeared likely to be repeated. Thus, Studies 3, 4, 5, and 6 were conducted without recording the concomitant EEG (resulting in 26, 21, 28, and 30 hits, respectively, all $p > .05$).

The investigator/subject thought it possible that this pattern of results (i.e., high performance in Study 1 and lower performance in subsequent studies) may have been due to a "beginner's luck" phenomenon, such as might be attributed to first-time gamblers. It is possible that when subjects perform guessing tasks like this for the first time—for which there is no rational way to select the target—they are not sure exactly how to go about it and thus may be more open to spontaneous thoughts, feelings, and hunches. As we become highly familiar with a task, we may tend to perform it in more set ways, and these may not be conducive to psi. In any event, this explanation jibes with the subject's conscious experiences during the task. It also accords well with the prevailing wisdom on how to succeed in a psi task.

In view of this, Study 7 was planned as another EEG experiment with the hypothesis that there would be a rebound of the (theoretical) beginner's luck effect. A rebound of performance was considered possible because nearly two years had elapsed since Study 6 was completed, and thus, it was reasoned, the task might again seem novel to the subject. Sadly, the results of Study 7 lent little support to the rebound hypothesis; the subject again scored 29 hits out of 100 trials.

However, viewing all seven studies as a whole, it may be seen that scores in 6 out of 7 were all in the predicted direction and that there may indeed have been a psi effect throughout the series of experiments, albeit with too little statistical power to detect it in (most of) the individual experiments. Combining the results of all seven studies, there were 199 hits out of 700 trials ($p = .021$; Cohen's $h = .09$). Because of the possibility that psi may have been operative throughout the entire series of experiments including Studies 2 and 7, the two studies for which EEG data were collected, we decided to undertake an analysis of the brainwave correlates of performance in these two studies.

EEG ANALYSES

For Studies 2 and 7, EEG electrodes were applied to the left and right anterior temporal and posterior temporal electrode sites, referred to linked ears. Trials containing eye or motion artifact were excluded, blind to trial accuracy. Pooling across both studies, we found 174 trials to be artifact-free. We compared hit trials and miss trials for EEG power in the theta (4–7 Hz), alpha (8–13 Hz), and beta (14–59 Hz) frequency bands during the final four seconds of the "impression period" of each trial.

Repeated-measures ANOVAs (Accuracy x Hemisphere x Topography) were performed on EEG power in each frequency band. Neither the main effect of Accuracy nor any interactions involving the Accuracy factor were significant in the analyses of theta or beta power (all $p > .05$). However, the Topography x Accuracy interaction was observed to be significant in the analysis of alpha power, ($F(1/172) = 11.90$, $p < .001$). Simple effects testing indicated that the interaction was due to greater alpha power for hits (147 μv^2) than for misses (136 μv^2) at the posterior sites, ($F(1/172) = 4.10$, $p < .05$), and reduced alpha power for hits (135 μv^2) than for misses (149 μv^2) at the anterior sites ($F(1/172) = 4.81$, $p < .03$). The hemisphere effect was also significant in the analysis of posterior alpha with power being greater at the left (144 μv^2) than at the right (134 μv^2) hemisphere ($F(1/172) = 5.45$, $p = .02$).

DISCUSSION

Analyses of the EEG data revealed that power in the alpha band differed for this subject between hit and miss trials. Changes in the power or dominant frequency of alpha activity have figured prominently in past EEG studies of psi. However, both psi-related increases in alpha activity (e.g., Palmer [*RIP* 1978, 135–138]) and psi-related decreases in alpha (e.g., Kelly & Lenz [*RIP* 1975, 22–25] and McDonough et al. [*JP* 1989, 182–201]) have been observed previously. Presently, the direction of difference in alpha power depended on scalp topography. Over the posterior recording sites we observed more alpha power for hit trials than for miss trials, whereas over the anterior recording sites we observed less alpha for hits than for misses. The meaning of the present interaction effect involving topography is not clear. It is noted, however, that the posterior sites, at which alpha power was presently observed to be greater for hits than for misses, are in close proximity to the occipital scalp where alpha is commonly recorded. Occipital alpha is thought to be associated with a state of restful alertness and is suppressed by intense mental effort.

During the impression period, the subject used a quasi-meditative technique characterized by low arousal and a marked reduction of higher cognitive processes, particularly those mediated by language, which is consistent with the alpha state described above. It is also consistent with the phenomenological experiences of successful psi subjects as described by White (*JASPR* 1964, 21–56). We may infer from the hit versus miss differences in alpha power that the alpha activity presently observed during hit trials reflected mental processes which were somehow conducive to the operation of psi. However, because self-reports of conscious experience were not made on a trial-by-trial basis, a one-to-one relationship between the contents of consciousness and performance accuracy cannot be established definitively for the present experiment. The greater alpha power observed at the left posterior recording site relative to the right posterior site is also consistent with the subject's self-reported reduction of verbal thought processes during the impression period.

On the other hand, the analyses of EEG power in the theta and beta frequency bands did not reveal any differences between hit and miss trials. Activity in the theta and beta frequency bands has also been previously linked to psi (e.g., Healy [*JASPR* 1986, 352–368], Don et al. [*RIP* 1987, 72–76; *RIP* 1988, 32–37], and McDonough et al. [*JP* 1989, 182–201]). Inconsistent findings from one study to another raise questions about the reliability of EEG effects similar to those that may be asked of other avenues of parapsychological inquiry. A skeptic might argue that nonsignificant or haphazard EEG effects are just the sort of thing one would expect by chance if psi were not operating in any of the experiments.

Although we cannot rule out the skeptical explanation, there are other credible reasons why EEG patterns might differ between studies. For one, if psi effects are only weakly elicited in the laboratory (which, at best, was presently the case for Studies 2 and 7), one should not expect strong consistencies in the EEG data between studies. A related possibility, well known to researchers in the field, is that even in an experiment with good behavioral evidence of psi, psi will almost certainly not be responsible for all of the hits observed in the experiment. Conceptually, we can partition the total number of hits observed in an experiment into chance hits and psi hits. Combining chance hits and psi hits in the same bin will necessarily dilute any EEG correlates of psi that may be present in those trials where psi was operative. To complicate matters more, it is generally acknowledged that some miss trials may also be due to psi, and such psi missing could further dilute or muddle any EEG comparisons between nominal hits and misses.

One final comment needs to be made on the issue of the reliability of EEG findings across studies. To be sure, it would be preferable to reduce the variance between studies due to weak psi effects or to the indiscriminate inclusion of chance hits and psi misses in the analyses. However, very high correlations between studies would not be desirable in the face of differing tasks, subject strategies, or levels of ability. EEG data may contribute to a fuller understanding of the psi process precisely because they are sensitive to fluctuations in the subject's mental states and activities.

Partial Replication of Single Subject Event-Related Potential Effects in a Psi Task*

Charles A. Warren†, Bruce E. McDonough, and Norman S. Don (Kairos Foundation and University of Illinois at Chicago). Address correspondence to Charles A. Warren, Ph.D., Kairos Foundation, 405 N. Wabash Ave., Suite 2201, Chicago, IL 60611

The study of physiological responses in psi tasks seems warranted in light of Beloff's (*JSPR* 1974, 403–420) observation that, if psi is largely an unconscious process, such responses may have an advantage over the typical verbal response, because they may circumvent the conscious decision-making process.

In a previous study by Warren, McDonough, & Don (*PA Proceedings* 1990, 334–460; *RIP* 1990, 71–76; *JP* 1992, 1–30), two out of five analyzed average, event-related brain potential (ERP) components, namely, the P100 and a negative slow wave (NSW), were found to be larger in response to target than nontarget stimulus images selected by a computer-based RNG and presented on a video display unit. The subject in both the previous and present studies, Malcolm Bessent, a well-known psychic, performed at chance level ($p > .05$) on a modified version of the ESPerciser™ task; this version delivered four foveal stimuli (visual angle: width/height was 2.42/2.99 degrees) sequentially (interstimulus interval = 2600 ms; stimulus duration = 330 ms), rather than simultaneously as with the standard version. (Thanks are in order to Charles Honorton for loan of the PsiLab II™ system and to George Hansen for assistance in system modification. Appreciation is also expressed to Mr. Bessent for his participation.) Two hundred self-paced trials (20 runs with 10 trials/run) were admin-

*This work was made possible by grant support from the Kairos Foundation.

istered with a trial rate of about one per 30 seconds; total testing time was about 100 minutes. The subject was encouraged to do as well as possible; experimenter attitudes were supportive. Rest breaks were permitted, and the subject was allowed to smoke at will.

The PsiLab II™ RNG output, tested using the default values of the system (100 samples with 20,000 trials/sample; Psychophysical Research Laboratories, *PsiLab II Users Manual.* Princeton, NJ: Psychophysical Research Laboratories, 1985), both prior to and following data collection, showed no significant departures from chance distributions ($p > .05$) in either Kolmogorov-Smirnov or the χ^2 tests, consistent with randomness.

Silver-silver chloride cup electrodes applied to scalp locations at the left and right hemisphere frontal (F_3, F_4), central (C_3, C_4), anterior temporal (T_3, T_4), posterior-temporal (T_5, T_6), and occipital (O_1, O_2) areas (international 10–20 system) were used. Reference electrodes for the electroencephalogram (EEG) were attached to the earlobes (A_1, A_2), balanced for resistance, and linked; a forehead ground was used. Eye blink and movement artifacts were indicated via the electrooculograph (EOG), recorded bipolar between the supra-orbital ridge and the outer canthus of the right eye. Impedances were 5 K ohms or less.

The EEG and EOG were amplified with custom-built MRA DC amplifiers (upper-end 6 dB roll-off at 60 Hz) and displayed on a Grass (Model 78D) polygraph strip-chart recorder. Physiological data and an event channel were stored for later off-line analysis on a Vetter (Model A) 16 channel FM tape recorder. Recorded calibration pulses permitted equalization of amplitude across channels.

EEG data were digitized at 125 Hz and low-pass filtered at 20 Hz, and trials were excluded which contained EOG greater than 30 uv or any amplifier blocking. One hundred fifty-nine trials were found to be artifact-free. The trial discard rate (21%), much smaller than in the previous experiment, is more in keeping with those seen in traditional published evoked-potential reports.

A pair of stimulus-related EEG epochs was selected from each trial, one target and one nontarget, in such a manner that a set of 159 target and 159 nontarget stimulus pairs was balanced for presentation order and subject calls (i.e., being picked by the subject as a target). In addition, the number of targets detected by the subject (hits) was matched by an equal number of nontargets picked by the subject (misses).

The 159 target stimuli were then divided into 10 sets of 12 stimuli and 3 sets of 13 stimuli each and averaged (sub-averages) to yield 13 average target waveforms. The stimuli sorted into these sets were aggregated according to their sequence of occurrence (e.g., the first series of 12 artifact-free target stimuli comprised the first set to be averaged; the second series, the second set; and so on). The corresponding set of nontargets was divided in the same way and averaged. The result was 13 average target waveforms with 13 matched, average nontarget ERP waveforms for each of 10 sites, a total of 260 average waveforms. Note that image background was also balanced between targets and nontargets since all four stimuli for each trial had the same background color.

Additional (ERP) data in response to target and nontarget stimulus categories were analyzed to verify whether the above-mentioned ERPs would show replicable or similar differences in an additional dataset. The data consisted of the 13 sets of target and 13 sets of nontarget averages, consisting of 10 sites per average for each ERP component. A repeated measures Stimulus (target, nontarget) X Hemisphere (left, right) X Topography (frontal, central, anterior-temporal, parietal, posterior-temporal, occipital) ANOVA showed that NSW (average of 400–500 ms of poststimulus onset activity relative to a 100 ms prestimulus baseline) gave rise to a significant Hemisphere X Stimulus interaction ($F[1/12] = 4.55$, $p < .054$). Greenhouse-Geisser corrections for repeated measures were applied to p values, where appropriate. Only Stimulus and Stimulus-related interactions were considered. Both the Stimulus ($df = 1/12$) and Stimulus X Topography ($df = 4/48$) tests were nonsignificant ($p \geq .4430$). A followup ANOVA showed the left hemisphere to have a slightly greater target than nontarget NSW ($F[1/12] = 2.40$, $p < .073$; one-tailed) but a nonsignificant Stimulus X Topography interaction ($F[4/48] = 1.09$, $p < .3443$). A followup ANOVA of the right hemisphere showed no significant effects (all $p \geq .6293$).

Contrary to the previous study, the same type of ANOVA on P100 provided no evidence for a target-nontarget difference (P100 measured as the average of 92–132 ms of baseline-referenced, poststimulus activity). All four tests were $p \geq .3907$. Consistent with the previous study, the same ANOVA showed Stimulus to have no significant effects or interactions for either N100 or P200 (all $p > .05$).

Reasons for Bessent's poor performance relate to possibly increased demands associated with the need to control eye movements and with the distractive effect of having electrodes attached (see McDonough, Warren, & Don, *JP,* 1990, 141–149).

Results for the NSW are weak but in keeping with an hypothesis of preconscious transformation of psi-received information (see Warren, McDonough, & Don, *JP,* 1992, 1–30). The subject could not overtly discriminate targets from nontargets, but the left hemisphere NSW was slightly greater for targets compared to nontarget stimuli. The same kinds of matching for target and nontarget categories for stimulus position across the four presentation positions and for subject calls (i.e., stimulus called target versus not called target) were in place as previously, thus ruling out imbalances in the foregoing variables as explanations for the current findings, as well as differences between the two experiments.

One possible reason for the failure to replicate the P100 effect is that the presence of an early, pre-P100 negativity to the target category may have confounded the accurate measure of P100 amplitude. Finding a left hemisphere locus of effect for NSW, instead of a more broadly distributed negativity, may have been due to the presence of a slow positivity for the nontarget category in the previous experiment, which was dramatically reduced to near baseline amplitude in the current experiment. Both the P100 and NSW findings here may have been due to a change in subject strategy, and future work should employ techniques such as principal components analysis to separate out the overlapping influences of other ERP components (see Donchin, E., & E. Heffley. "Multivariate analysis of event-related potential data: A tutorial review." In *Multidisciplinary Perspectives in Event-related Brain Potential Research* (EPA-600/9-77043), edited by D. Otto. Washington, DC: U.S. Government Printing Office, 1978). Future studies could also perhaps discover phenomenological "handles" that would mark the occurrence of such strategy changes. These could be developed by asking subjects appropriate questions during testing, such as questions designed to get at the subject's use of certain cognitive operations in response to each stimulus. Such operations as mental manipulation of visual images, scanning of internal memory stores (both known to be associated with slow wave negative ERPs), and rule-learning and memory-encoding operations (known to be associated with slow wave positive ERPs) could be explored (see review by Ruchkin, Johnson, Mahaffey, & Sutton, *Psychophysiology,* 25, 1988, 339–353).

GANZFELD STUDIES

The Ganzfeld Novice: Four Predictors of Initial ESP Performance

*Charles Honorton (University of Edinburgh, 7 George Square, Edinburgh, Scotland EH8 9JZ)**

One of the goals of the PRL automated ganzfeld research was to identify characteristics associated with successful ESP ganzfeld performance by previously inexperienced participants (novices). Two hundred and six participants each contributed to one of five novice series in the PRL automated ganzfeld project (Honorton et al., *JP,* 1990, 99–139). Most participants completed Form F of the Myers–Briggs Type Indicator (MBTI) and a 55-item demographic survey (the Participant Information Form, or PIF).

Three previous reports have already described aspects of the PRL novice research. The relationship between novice ESP Ganzfeld performance and extraversion was presented by Honorton, Ferrari, & Bem (*RIP* 1990, 35–38). Schlitz and Honorton (*JASPR* 1992, 83–98) described a subset of the PRL novice data involving performing artists from the Julliard School. At the 1986 PA Convention, Honorton and Schechter (*RIP* 1986, 36–39) presented an exploratory analysis of performance correlates for the first two PRL novice series (Series 101–102; hereafter designated

*I wish to express my gratitude to the American Society for Psychical Research, Inc., the Parapsychology Foundation, Inc., and the Koestler Chair of Parapsychology, University of Edinburgh, for their financial support.

PRL-1), suggesting that initial ganzfeld ESP performance was positively and significantly related to self-reports of personal psi experiences, Feeling/Perception (FP) preferences on the MBTI, and prior participation in nonganzfeld psi experiments. A positive but nonsignificant tendency for better performance among participants reporting involvement with mental disciplines such as meditation was also found.

Broughton, Kanthamani, and Khilji (*RIP* 1989, 32–35) successfully replicated the ESP/FP finding in an independent study at FRNM. In this paper, the PRL-1 findings will be compared with those in the later PRL novice series (Series 103–105; hereafter designated PRL-2) and the FRNM series to estimate the overall magnitude and consistency of the four predictors.

Following presentation of the Honorton & Schechter report, an effort was made to obtain Myers-Briggs data from PRL-1 participants who had not originally taken the MBTI, and ten participants kindly cooperated. Their data are included in the PRL-1 totals in this report. The PRL-1 findings summarized in this paper also reflect corrections for a number of data entry errors in the Honorton & Schechter report that were discovered during a data audit of the entire PRL automated ganzfeld database prior to publication of Honorton et al. (*JP,* 1990, 99–139).

The PRL novices included 121 women and 85 men (total n=206). PIF data were available for 195 and MBTI data were available for 190 subjects. These participants tended to be mature, moderately well educated, and, perhaps most prominently, strong believers in psi. Of the MBTI scales, only the SN (Sensing/Intuition) MBTI Scale was unusual; PRL novices tended strongly toward the Intuition side of the scale. The two PRL studies differed significantly on responses to four PIF items: PRL-2 subjects were younger and reported fewer personal psi experiences, a lower frequency of coincidences, and, curiously, stronger belief in luck than those in PRL-1. These differences, with the exception of luck, are due to the Julliard students in PRL-2 (Schlitz & Honorton, *JASPR,* 1990, 91).

PIF data were available for 108 of the 120 FRNM novices. The PRL and FRNM novices also differed significantly in four areas: PRL novices were older, had a higher mean on the MBTI SN Scale, reported more personal psi experiences, and reported more involvement in mental disciplines.

PSI PERFORMANCE

Thirty percent of the PRL-1 novices, 35% of the PRL-2 novices, and 28% of the FRNM novices correctly identified their targets. The three studies combined yielded a success rate of 31% (n=326, p=.012, z=2.25). The effect size (Cohen's h; J. Cohen, *Statistical Power Analysis for the Behavioral Sciences,* Academic Press, 1977) was .13, and the 95% confidence interval (CI) from 26% to 36%. The effect sizes of the three studies are quite consistent (χ^2=2.39, 2 df, p=.30).

REPORTED PERSONAL PSI EXPERIENCES

Following standard definitions of the four basic psi types, PIF item 14 asks: "If you have had experiences which you thought involved psi, which of the following do you feel you have experienced (please check)." One point was given for each of the checked items (telepathy, clairvoyance, precognition, psychokinesis), and their sum constituted the psi experiences predictor.

Honorton & Schechter (*RIP* 1986, 36–39) found a significant positive correlation between the number of types of psi experiences and psi ganzfeld performance (session z-scores but *not* direct hits) for participants in PRL-1. Recast as a one-way ANOVA using linear contrasts, F=7.41, 1/86 df, p=.0039, r=.28. In PRL-2, F=2.86, 1/99 df, p=.047, r=.17. When the Julliard students (who reported significantly fewer types of personal psi experiences) are excluded from the PRL-2 analysis, the relationship between psi performance and psi experiences is nearly identical to that obtained in PRL-1 (F=6.29, 1/79 df, p=.007, r=.27). Combining the two PRL studies gives F=9.37, 1/190 df, p=.0015, r=.24.

Broughton et al. reported a modified analysis between self reports of psi experiences (coded as "none" or "some") and direct hits. FRNM novices reporting "some" personal psi experiences were non-significantly more successful than those reporting "none." If we combine the three studies by using the FRNM psi experience definition as the predictor and direct hits as the outcome measure, the combined difference between participants who did and did not report personal psi experiences is not significant; but significant ESP performance was obtained only by those who reported psi experiences (n=272, 33% hits, p=.0015, z=2.96, h=.19).

The stronger association in the PRL data between number of types of personal psi experiences and session z-scores (rather than direct hits) indicates that persons who believe they have had a variety of psi experiences are more adept at identifying partial correspondences between their ganzfeld experience and targets in the judging pool. Further research is needed to determine whether this is better explained by motivational or cognitive factors.

PRIOR PSI TESTING

Sixteen percent of the PRL novices and thirteen percent of the FRNM novices had participated in other types of formal laboratory testing of psi phenomena. In PRL-1, novices who had previously participated in other, nonganzfeld psi experiments obtained a hit rate of 50%, versus 26% for those with no prior psi testing experiences; in PRL-2, the hit rates were 67% for those with prior psi testing experience, 32% for those with none. The FRNM study showed a slight reversal of the trend, with a hit rate of 29% for participants with prior experience and 31% for those with no prior experience. The overall effect of prior psi testing is significant (though clearly further research will be needed to assess the cross-laboratory generality of this finding): 48% of those with prior psi testing obtained hits (n=46, p=.0021, z=2.87, h=.47, 95% CI from 35% to 60%), compared with 30% of those with no prior testing (n=258, p=.053, z=1.62, h=.11, 95% CI from 25% to 35%).

MYERS-BRIGGS FEELING-PERCEPTION

Following MBTI convention, participants were classified as FP if their continuous scores on the TF and JP Scales were both above 100. In PRL-1, 50% of the MBTI FP participants obtained hits, compared to 18% of those classified non-FP. In PRL-2, the success rates for FP and non-FP participants were 36% and 35%, and in the FRNM series the hit rates were 40% and 25%. When we combined the three studies, the FP subjects show significant performance, with a hit rate of 42% (n=127, p=.000064, z=3.83, h=.38, 95% CI from 34% to 49%). Non-FP subjects show non-significant overall performance, with a hit rate of 27% (n=164, p=.33, z=0.44, h=.035, 95% CI from 20% to 34%). The difference in overall success rates between FP and non-FP subjects is also significant (z=2.69, p=.0036). Using meta-analytic techniques for combining correlations, the mean weighted phi is .16 (95% CI from .05 to .27).

MENTAL DISCIPLINES

In PRL-1, hits were obtained by 36% of those reporting some use of mental disciplines, such as meditation, biofeedback, hypnosis, or relaxation exercises, and by 16% of those with no involvement with such mental disciplines. In PRL-2, hits were obtained by 33% of participants with mental disciplines experience and by 48% of the non-practitioners; the latter outcome is largely due to the Julliard non-practitioners, who had a success rate of 71%. In the FRNM study, practitioners had a 34% hit rate, compared to 26% for the non-practitioners.

When we combine the three studies, 34% of the practitioners were successful (n=220, p=.0025, z=2.81, h=.20, 95% CI from 28% to 40%), compared to 28% of the nonpractitioners (n=83, p=.37, z=0.33, h=.07, 95% CI from 20% to 38%). The difference between practitioners and nonpractitioners is nonsignificant (z=0.96, p=.168, phi=.06). When the Julliard students are excluded, the overall difference increases to z=1.45, p=.086, phi=.09.

COMBINED-PREDICTOR MODELS

There are significant intercorrelations among three of the predictors in the combined PRL data: reported psi experiences, FP, and mental disciplines. The strongest correlation is between FP and psi experiences (r=.226, 185 df, p=.002). The correlation between FP and involvement with mental disciplines is .184 (p=.012). Mental disciplines and psi experiences show a correlation of .179 (193 df, p=.012).

The Four-Predictor Model

Due to the small number of participants with prior psi testing experiences, only fifteen participants in the three studies satisfied the four-predictor model (reported psi experiences, FP, mental disciplines, and prior testing). Nevertheless, the results are rather striking. Eleven of the fifteen participants correctly identified their targets (73% hits, weighted z=3.35, p=.00041). The overall effect size (h weighted by sample size) is 1.03.

The Three-Predictor Model

In the three studies combined, 99 novice participants satisfied the three-predictor model (reported psi experiences, FP, and involvement with mental disciplines). They achieved an overall success rate of 43% (p=.00014, z=3.64, h=.39, 95% CI from 35% to 52%). The 191 participants not satisfying the model obtained a success rate of 28% (p=.179, z=0.92, h=.07, 95% CI from 20% to 38%). The difference between the two groups is significant (p=.0041, z=2.64), and the phi coefficient is .15.

DISCUSSION

Although none of the predictors individually differentiated successful versus unsuccessful performance in each of the three studies, significant hitting across all three studies was limited to participants meeting the criteria for each of the individual predictors. From a purely pragmatic point of view, if we were advising new ganzfeld investigators how to maximize the likelihood of successful psi ganzfeld performance in their studies, we would recommend that they recruit participants who conform to as many of three factors—plus extraversion (Honorton, Ferrari, & Bem, *RIP* 1990, 35–38)—as possible.

More detailed and systematic refinement of the four predictors used in these studies would probably be more useful at this stage than further direct replications involving the current implementation. For example, a more refined measure of participants' putative psi experiences might enable investigators to test whether the modality of participants' personal psi experiences is systematically (and perhaps differentially) related to performance in different types of experimental task.

An Experiment in Ganzfeld and Dreams: A Further Confirmation

H. Kanthamani† *and Richard S. Broughton (Institute for Parapsychology, 402 N. Buchanan Blvd., Durham, NC 27701–1728)*

The present experiment is a third study in the series, comparing ganzfeld and dream conditions in a free-response ESP test. The earlier two studies (Kanthamani et al., *RIP* 1989, 84–89; Kanthamani & Khilji, *RIP* 1990, 56–61) had showed a significant difference between ganzfeld and dream scores, favoring the dream condition. Whereas the first experiment was with a single subject, the second included ten subjects, but the findings needed to be replicated on a larger sample size. The present study was conducted for this purpose. The main objective, once again, was to see whether the dream condition would continue to be better than ganzfeld in eliciting psi information. From this standpoint, this study serves as another replication attempt, along with certain extensions.

Twenty volunteer subjects (9 males and 11 females) participated in this experiment. As before, a within-subject design was followed, with the same subjects doing both ganzfeld and dream conditions in each trial. They had ganzfeld as the first condition and dream as second (G-D order) in their first trial; the reverse pattern (D-G order) was followed in the second trial. Thus, the two orders of presentation were not counterbalanced but were deliberately kept constant for all the subjects, mainly because we felt that was the optimum situation, based on some of our results in the previous study. In all there were 40 ganzfeld sessions, 20 followed by dream and 20 preceded by dream. Likewise, there were 40 dream sessions, between the two presentations, equally contributed by the 20 subjects.

TARGETS

The target pool consisted of 50 sets of 4 pictures, which represents the standard pool (constructed by Kanthamani & Khilji in 1986) that has been in use for ganzfeld and other free-response work at the Institute. The individual pictures in each set remained in sealed envelopes, along with a judging packet (also sealed) consisting of a duplicate set of the four pictures. For each trial, RSB used a computer pseudorandom number generator to select a target. At the completion of the trial, the target set was reassembled and inserted back into the pool.

PROCEDURE

The essential procedure was identical to the previous studies. It can be described under three main parts, covering the ganzfeld, dream, and judging tasks. The ganzfeld session took place at the Institute, and the subject was instructed to bring a report of the night's dreams collected at home for evaluation the following day. The ganzfeld procedure included the standard technique, with ping-pong ball halves affixed over the eyes, red light stimulation, and pink noise with relaxation instructions. However, it was different from the standard procedure in the following respects: (1) Instead of the usual GESP method, a clairvoyance technique was used, because this was more convenient; (2) The judging process was conducted only when both ganzfeld and dream conditions were completed in a given trial; (3) Both the subject and the experimenter(s) acted as judges.

For the dream part of the trial, the target remained at the Institute, and the subjects attempted to get target information through their nocturnal dreams. They were told to focus on it before going to bed and to give a suggestion to themselves that the target information would enter into their dreams. They were also instructed to keep some writing material nearby when they went to bed, so that they could jot down some impressions of the dreams if they woke up in the middle of sleep. As soon as they got up in the morning, they were required to write down a full report of their dreams, which they brought to the lab for evaluation.

As noted above, the judging of the protocols took place when both parts of the trial—ganzfeld & dream conditions—were completed. Each judge separately compared each mentation against the four pictures from the judging pool and recorded the correspondences in terms of ranks (1–4) and ratings (0–99). The protocol from the condition presented first was evaluated before the second condition protocol.

RESULTS

Based on our earlier work, a significant difference in the scoring rate between ganzfeld and dream conditions was predicted, in favor of dreaming. Since the pooled scores of the different judges formed the main measure in the earlier studies, the same was done here also. This was done by combining the judges' respective ratings on different targets in a set and transforming them into z scores, using Stanford's 1974 method (Stanford & Sargent, *JASPR,* 1983, 319–326). The z scores were obtained separately for each condition in a trial. This produced 80 z scores, between the two conditions and the two orders of presentation, which were then analyzed by a two-way ANOVA. The main hypothesis relating to the ganzfeld-dream differences was being tested by the first main effect; any possible order effects were to be revealed in the second main effect.

The ANOVA showed that the main effect relating to the conditions yielded an F=3.699, 1, 19 df, p=.035, confirming our hypothesis that there is a significant difference between ganzfeld and dream conditions. The mean z scores were .089 for the ganzfeld and .331 for the dream. As predicted, the scoring difference is in favor of the dream condition, and it was strongly significant (t=3.521, 19 df, p=.00233). The ganzfeld scores were also above chance, but not significant (t=.607). The second main effect relating to the presentation order yielded an F=2.065, 1, 19 df, p=.084, suggesting a weak order effect. The interaction, however, was not significant (F=0.031).

Secondary analyses included separately carrying out a similar two-way ANOVA for each judge. The results showed that the condition effect was independently significant by both the percipients' judging (F=3.135, 1, 19 df, p=.047) and the experimenter's judging (F=3.022, 1, 19 df, p=.049). The dream scores were significantly higher than MCE for the percipients (Mz=.353, $p < .05$) but not for the experimenter (Mz=.234, ns). The ganzfeld scores in both cases were nonsignificant (Mz=.081 and .043 for the percipients and the experimenter, respectively). However, a strong presentation-order effect emerged only by the experimenter's judging (F=6.648, 1, 19 df, $p < .01$), indicating that the condition that was presented and judged second (Mz=.280) produced substantially higher results than the condition presented first (Mz=.028). There was no evidence of an interaction effect in either case.

DISCUSSION

From the results presented above, it is clear that there is an overall condition effect, representing a significant difference between ganzfeld and dreaming. Secondly, the difference is in favor of dreaming, which is independently significant. These results fully confirm our hypothesis and serve as a successful replication of the previous experiment.

The finding that dreaming is more successful than ganzfeld is interesting. In this experiment, the direct hit rate for the subjects in ganzfeld was 33%; however, in the dreaming condition it was still higher at 38%. Perhaps the naturally occurring altered state functions more efficiently in gaining psi information than the laboratory-induced altered state condition.

The two factors observed in these data—viz., the superiority of dreams over ganzfeld and the tendency for the condition judged second to be better—combined to make the GD order (where ganzfeld came first and dream second) the more successful in highlighting the ganzfeld-dream difference, thereby following the pattern of the previous experiment. Initially, this was interpreted as something like the Poetzl effect, wherein the "psi stimulus" of the ganzfeld was incorporated into later dreams; however, the results of this experiment suggest that other factors may be involved. It remains for future research to determine whether the effect is a result of stimulus carry-over or judging carry-over.

PK STUDIES

Beyond Belief: Exploring Interactions Among Mind, Body, and Environment

Dean I. Radin (Institute for Advanced Study at Waltham, 253 Bishops Forest Drive, Waltham, MA 02254)

Thirteen PK experiments conducted by four participants over five months explored interactions among mind, body, and aspects of the local, global, and cosmic environment. Specifically, the effect of focused mental intention on the level of background ionizing radiation was studied, along with physiological correlates of success at the assigned task and possible influences of the local magnetic field, planetary geomagnetic field, and sunspot number. The basic hypothesis was that the level of background radiation would drop when a person willed it to drop.

The experiment continuously measured background radiation at a fixed site using a computer-monitored Geiger counter. At random intervals, a tone sounded; this signalled the subject to mentally "intend" or will the count rate returned by the Geiger counter rate to be as low as possible. The mean of mentally "treated" samples was compared to the mean of control samples (nontreated) to see whether they differed in accordance with the instructions.

Outcomes of seven experiments were in the predicted direction; three were significant at $p < .05$. Time series analyses of experiments supporting the mental influence hypothesis revealed that the background radiation level dropped in accordance with task instructions only during randomly assigned treatment periods. Adjacent control periods showed no similar drops. Background radiation levels unexpectedly rebounded above the background rate 20 seconds after the treatment period.

One of the 13 experiments investigated physiological and local magnetic field correlates of success at the PK task. Results of this experiment indicated that blood volume, pulse, heart rate, and the local magnetic field responded differentially when radiation counts were low. Analysis of geophysical and geocosmic factors at the time of the experiments indicated that the baseline level of background radiation was related to success in the experiment, with more successful experiments occurring during days of lower ambient radiation. In addition, successful experiments tended to occur on days in which the sun displayed lower sunspot numbers.

The overall mean effect size for the PK task was less than one standard error from a chance effect, but mean effect size adjusted by simply taking into account the change in sunspot number from the day before to the day of each experiment was more than two standard errors from chance.

The present series of experiments suggest that

(a) Mental intention can sometimes influence naturally occurring levels of background ionizing radiation, but this influence is modulated in some complex way with changes in sunspot numbers and with the baseline level of background radiation. An unexpected rise in radiation observed after successful mental "treatment" periods suggests that PK in this task acted as a local entropy-reduction mechanism which acts in time. That is, small declines in entropy at a specific site seem to be balanced out later in time by slight increases in entropy at the same site.

(b) The physiological state of the person applying the mental intention, as well as the local magnetic field, may also correlate with PK success. Future experiments in this realm may be more productive if ensembles of variables are monitored, rather than if we assume that the normal experimental controls are adequate to exclude all external influences. At this early stage in understanding subtle mind-body interactions, we simply do not know what variables are important to include or exclude. For example, one can monitor but not control solar flares and sunspot numbers, and it may turn out that these and other solar effects significantly influence success at these tasks.

In future studies specifically involving background radiation, I recommend that multiple radiation detectors be employed. A two-

or three-dimensional array of Geiger counters, for example, with one particular Geiger counter defined as the mental target, may help determine the mechanisms of the mental influence. For example, during successful trials, is (a) radiation deflected away from the target, (b) radiation holistically lowered in the local vicinity, or (c) the sensitivity threshold of the target Geiger counter raised? Many other interesting possibilities can be explored.

PSI AND PSYCHOLOGICAL FACTORS

Sensation Seeking, Drug Use, and Reported Paranormal Beliefs and Experiences

V.K. Kumar† (Department of Psychology, West Chester University, West Chester, PA 19383), Ronald J. Pekala (Coatesville VA Medical Center), and James Cummings (West Chester University)

The present study examined the relationship between reported paranormal beliefs and experiences and sensation-seeking behavior. Specifically, we asked whether paranormal beliefs and experiences are related to the General Sensation Seeking trait or more narrowly to one or more of its four traits (Boredom Susceptibility, Disinhibition, Thrill and Adventure Seeking, and Experience Seeking). Since males and females seem to differ not only with respect to their sensation-seeking behavior but also in the extent of their paranormal beliefs and experiences, another purpose of the present study was to see whether gender moderated the relationship between sensation seeking and reported paranormal beliefs and experiences.

Furthermore, since sensation-seeking behavior is related to experimenting with drugs or other mind-altering substances, we decided to include "drug use" as a factor in the analyses. Thus, it was possible to (a) relate "drug use" to the sensation-seeking scale to replicate earlier work and (b) examine whether gender, drug use, and sensation seeking interacted in their relationship with paranormal beliefs and experiences.

METHOD

Subjects and Instruments

Subjects were 229 male and 345 female Introductory Psychology students.

Paranormal beliefs and experiences were measured using the Mental Experience Inventory (MEI), which comprises seven subscales (Pekala, Kumar, & Cummings, *JASPR,* 1992, 136–150): Belief in Psi-related and Unusual Events (e.g., life after death, witchcraft/sorcery); Paranormal and Unusual Experiences (e.g., trying channelling, using an Ouija board); Sense of Being High (feeling intense emotions); Daydreaming (fantasizing); Sense of Mental Potency (tremendous power of mind); Introspection (analyzing self); and Altered Consciousness (getting lost in one's own thoughts).

Sensation seeking was measured by Zuckerman, Eysenck, and Eysenck's Sensation Seeking Scale, (*Journal of Consulting and Clinical Psychology,* 1978, 139–149) which yields a measure of General Sensation Seeking (General SS) and four subscale measures: Boredom Susceptibility (BS—avoiding repetitive activities); Disinhibition (DIS—engaging in uninhibited social activities); Thrill and Adventure Seeking (TAS—engaging in novel, dangerous, high speed activities); and Experience Seeking (ES—seeking novel experiences through drugs, music, travel, and art).

RESULTS

Frequency distributions of the General SS scale and its four subscales were examined to identify a group of low and a group of high sensation seekers on each of the (sub)scales. The cutoff scores were selected to achieve a balance between maximal separation of the low from the high groups and to have an adequate number of subjects (about 9–10) in each cell to perform a multivariate, three-factor (2x2x2) analysis of variance. The factors were: Sensation Seeking (high, low); Gender (male, female); and Use of a Mind-Altering Substance or Drug Use (yes, no). The MEI subscales served as the dependent variables. A significance level of .01 was used for both multivariate and further univariate analyses of variance.

General Sensation Seeking Scale

The "SS by Drug Use" frequency breakdown showed that among males, a significantly higher percentage of high (71%), relative to low (17%), General SS subjects reported having tried drugs ($z=5.46$, $p < .01$). For female subjects, 21% of the low and 66% of the high General SS subjects reported having tried drugs ($z=5.19$, $p < .01$).

The three factor MANOVA (with MEI subscales as dependent variables) revealed none of the two- and three-way interactions to be significant. Regarding main effects, Gender was not significant, but Drug Use and Sensation Seeking were significant ($F[7,256]=6.15$ and 6.88, $p=.000$).

Further univariate analyses on Drug Use and Sensation Seeking revealed that (a) drug users, relative to nonusers, endorsed a significantly higher number of items on Psi-Related Beliefs, Paranormal Experiences, Introspection, and Altered Consciousness subscales; and (b) high, relative to low, General SS subjects endorsed a significantly higher number of items on the Psi-Related Belief, Paranormal Experiences, Daydreaming, and Introspection subscales.

Sensation Seeking Subscale Analyses

1. *Boredom Susceptibility Subscale* (BS). The "BS by Drug Use" frequency breakdown showed that among males, a significantly higher percentage of high (53%), relative to low (32%), BS subjects reported having tried drugs ($z=2.25$, $p < .05$). The difference between the high (46%) and low BS female (31%) subjects was not significant ($z=1.79$, $p > .05$).

MANOVA revealed no significant two- and three-way interactions. The main effects of Gender and Drug Use were significant ($F [7,283]=3.19$ and 7.67, $p=.003$ and .000 respectively). The main effect of BS was not significant ($F [7,283]=1.55$, $p=.151$).

Further univariate analyses on Gender and Drug Use revealed that (a) males scored significantly higher than females on the

Sense of Mental Potency subscale and (b) drug users scored significantly higher than nonusers on Paranormal Beliefs, Paranormal Experiences, Daydreaming, Introspection, and Altered Consciousness.

2. *Disinhibition Subscale* (DIS). The "DIS by Drug Use" frequency breakdown showed that among males, a significantly higher percentage of high (61%), relative to low (21%), DIS subjects had tried drugs (z=4.23, p < .01). Among females, 20% of the lows and 63% of the highs (z=5.49, p < .01) reported having tried drugs.

MANOVA revealed no significant two- and three-way interactions. Gender and DIS main effects were not significant, but Drug Use was significant (F [7,278]=7.68, p=.000).

Further univariate analyses on Drug Use showed that drug users, relative to nonusers, scored significantly higher on Psi-Related Beliefs, Paranormal Experiences, and Introspection subscales.

3. *Thrill and Adventure Seeking Subscale* (TAS). The "TAS by Drug Use" frequency breakdown showed that among males, a significantly higher percentage of the high (54%), relative to low (26%), TAS subjects reported having tried drugs (z=3.54, p < .01). Among females, 34% of the lows and 42% of the highs reported having tried drugs (z=1.18, p > .05).

MANOVA revealed no significant two- and three-way interactions. The main effects for Gender, Drug Use, and TAS were all significant (F [7,329] ratios ranged between 7.53 and 9.17, p=.000 in all cases).

Further univariate analyses on these main effects showed that (a) females scored significantly higher than males on Daydreaming and Introspection; (b) drug users scored significantly higher than nonusers on Psi-Related Beliefs, Paranormal Experiences, Daydreaming, Introspection, and Altered Consciousness; and (c) high, relative to low, TAS subjects scored significantly higher on Psi-Related Beliefs, Paranormal Experiences, Sense of Being High, Daydreaming, Sense of Potency, and Introspection.

4. *Experience Seeking Subscale* (ES). The "ES by Drug Use" frequency breakdown showed that among males, a significantly higher percentage of the high (79%), relative to low (22%), ES subjects reported trying drugs (z=5.94, p < .01). Among females,

19% of the lows and 58% of the highs ($z=5.20$, $p < .01$) reported drug use.

MANOVA revealed no significant two- and three-way interactions. The main effects of Gender, Drug Use, and ES were all significant; $F(7,270)$ ratios respectively were 2.05 ($p=.004$), 7.89 ($p=.000$), and 9.62 ($p=.000$).

A marginally significant univariate Gender main effect was obtained for the subscale Sense of Mental Potency ($F[1,276]=5.78$, $p=.017$). The proportion of items endorsed by males and females were .67 and .54. Other univariate Gender main effects were not significant.

Further univariate analyses on Drug Use showed that drug users, relative to nonusers, scored significantly higher on Psi-Related Beliefs, Paranormal Experiences, Daydreaming, and Introspection.

Significant univariate main effects for ES were obtained for all subscales. On each subscale, high, relative to low, ES subjects endorsed a higher number of items.

SUMMARY AND CONCLUSIONS

There were five main findings: (a) high, relative to low, sensation seekers (particularly males) were more likely to report using drugs; (b) drug users, relative to nonusers, expressed stronger belief in psi-related phenomena, reported more frequent paranormal experiences, and tended to be more introspective; (c) high, relative to low, sensation seekers (General SS, TAS, and ES, but not BS and DIS) expressed stronger belief in psi-related phenomena, reported more frequent paranormal experiences, and tended to be more introspective; (d) males and females did not differ in their extent of paranormal beliefs and experiences; and (e) the two- and three-way interactions were not significant in any of the analyses.

The results suggest that using drugs and believing in and experiencing paranormal phenomena are both types of sensation-seeking behaviors that may arise from the tendencies to seek thrill and adventure (TAS) and new and varied experiences (ES). Whereas boredom susceptibility (BS) and disinhibition (DIS) appear to be of significance in drug use, they are less so in paranormal beliefs and experiences. It is possible that drug use and interest in paranormal phenomena may occur more for introspective reasons (personal enlightenment, discovering one's abilities) than for meeting social needs or yielding to peer pressure. It appears that the Thrill and Adventure Seeking and Experience Seeking sensation seekers are more prone than the other two types (Boredom Susceptibility and Disinhibition) to be drawn to ideas that go beyond current scientific verification (Zuckerman, *Sensation Seeking: Beyond the Optimal Level of Arousal,* Lawrence Erlbaum, 1979). High, as opposed to low, BS and TAS sensation seekers may also use drugs merely to alter consciousness (to feel different, let go of one's mind). Finally, drug use and gender did not moderate the relationship between sensation seeking and belief and/or experiences in paranormal phenomena, since the two- and three-way interactions were not significant in any of the analyses.

The Effect of a Threatening Subliminal Stimulus on the Perceptual ESP Test: A Partial Replication

John Palmer (Institute for Parapsychology, 402 N. Buchanan Blvd., Durham, NC 27701-1728)

In research reported at last year's convention (*RIP* 1991), 80 Dutch subjects, most participating for college course credit, completed two 20-trial GESP runs using the Perceptual ESP Test (PET), a procedure in which responses are determined by random eye fixations on one of an array of target symbols flashed to the subject. Immediately prior to each flash of this target slide, a slide from the Defense Mechanism Test (DMT) including the countenance of either an angry (threatening) or smiling man was flashed subliminally. Two significant post-hoc findings of note were reported: (1) tight variance of ESP scores in the total sample and (2) psi-hitting on targets called six or fewer times per run, a confirmation of Stanford's response-bias hypothesis. Two explanations were proposed for the tight variance: (1) subconscious threat induced by both DMT slides (J.P. suggested retrospectively that the smiling "control" face might have been interpreted as lecherous) and (2) a negative reaction to a visual field manipulation requiring subjects to wear uncomfortable goggles and use a chin rest.

The present experiment was designed to see if these findings could be replicated with an American sample and to eliminate the second interpretation by discarding the visual field manipulation and by restricting participation to subjects interested in the study. Letter targets were replaced by carets (Vs) in an effort to reduce response biases. Changes based on economic considerations included eliminating the control condition and the agent and using a single 40-trial run. A "pass" option was also introduced into the PET.

METHOD

Subjects

Forty volunteer subjects were recruited primarily from the Duke University student population. They were told only that the experiment involved ESP and subliminal perception.

Targets

The target slide consisted of 28 carets (seven pointing up, down, left, and right) distributed uniformly over the area of the slide on a grayish background. A total sequence of 1,760 unobserved targets (including 160 reserves) was generated by a pseudorandom algorithm using Turbo-Basic software and seeded by the number of microseconds between two keyboard button presses on the computer. The selection process was repeated two or three times until statistical tests for singlet bias (total and summed over runs), doublet bias, and bias for XX doublets all gave outcomes equivalent to less than ±1.5 standard deviations from chance expectancy.

Layout and Equipment

The subject and the experimenter (J.P.) were located in adjacent rooms separated by a closed door and a glass window. In the experimenter's room, an IBM-PC-type microcomputer controlled a two-channel Gerbrands projection tachistoscope that flashed stimuli through a polarizing filter, through the window, and finally through a 12 × 14 cm, 12 mm-thick ground glass screen mounted in the subject's room. The subject sat in a reclining chair 2 m in front of the screen.

PET Procedure

The subject initiated each trial by pressing a green button on a response box and looking up at the center of the screen. After two seconds, the threatening DMT slide (an angry man looking over the right shoulder of a seated boy) was flashed for 10 ms with the room dimly lit. Subjects were then to allow their eyes to wander around the screen. Two seconds later the target slide was flashed for 150 ms. Subjects either recorded their target choice by pressing the appropriate white button on the response box or passed by pressing a red "pass" button that caused the trial to recycle with the target unchanged.

General Procedure

Most subjects were first given a tour of the Institute by staff member Kathy Dalton. J.P. then took the subject to the test room. After a few minutes of friendly conversation, he explained the test procedure, identifying the DMT slide only as a subliminal picture. After signing the consent form and completing two questionnaires measuring dissociative experiences and anxiety respectively, the subject was guided through ten PET warm-up trials by J.P. A neutral slide of two boys standing against a wall was used in place of the DMT slide for these warm-up trials.

Following the official 40-trial PET, the subject completed a short rating scale concerning reactions to the procedure and was then asked by J. P. to describe the neutral and DMT slides at full exposure. Finally, J. P. recovered the printout of the ESP results, showed it to the subject, and completed the debriefing.

RESULTS

The rating scale indicated that during the PET itself no subjects were able to identify the threatening man as a living being and none attributed negative affect to the picture. During full exposure, 23 of the 37 subjects who gave answers (62%) used negative words such as angry or scary to describe the man.

The mean ESP hit total for the 40 subjects was 10.33 (SD=2.28, CR=0.90, p=n.s.). MCE was 10.

As predicted, the mean variance around MCE (5.31) was lower than the expected value of 7.5, but it was not quite significant (CR= − 1.46, p=.072, one-tailed). A secondary goodness-of-fit test also approached significance (χ^2=9.05, 4 df, p=.060, two-tailed), and it revealed that the tight variance trend was attributable to an excess of scores ±1 deviation from MCE. This same pattern occurred in the Dutch experiment. If the two experiments are pooled and the ±1 category is compared to the other categories combined, the result, although transparently post-hoc, is highly significant, (χ^2=15.09, 1 df, p=1 x 10^{-4}).

Of the 108 counterbias responses (called six times or fewer), 31 (28.7%) were hits (CR=0.78, p=n.s.). A goodness-of-fit test revealed that the degree of call imbalancing was slightly greater than in the Dutch study, and so the failure to replicate cannot be attributed to the carets reducing counterbias responding.

Exploratory analyses revealed that ratings on a two-item subscale measuring reported difficulty in coping with the caret test correlated positively with subject-based ESP variance scores (r(s)= +.41, 38 df, p=.008). This means that the tight variance was attributable primarily to subjects giving low difficulty ratings. The difficulty subscale almost correlated significantly with ESP hits as well (r(s)= +.28, 38 df, p=.083). The correlation between the hits and variance scores was −.01. The overall mean score on the 100-point difficulty subscale was 26.23.

Nineteen subjects indicated that at least once during the session they had a mental image of one of the targets before the target slide flashed. These subjects had a mean ESP hit total of 11.21 (SD=1.87) [t = 2.82, 18 df, p = .011]. They also scored higher (t=2.49, 38 df, p=.017) than the other 21 subjects, whose mean was 9.52 (SD=2.36). Fifteen of the 19 subjects who answered the imagery question affirmatively gave estimates ranging from 0 to 30% as to the percentage of total trials in which such images determined their ESP response. This measure correlated positively with ESP hits (r(s)= +.66, 13 df, p=.008), as well as ESP variance scores (r(s)= +.70, 13 df, p=.004). For this subsample, the ESP hit and variance scores were highly correlated (r(p)= +.71, 13 df, p=.003).

DISCUSSION

The tight variance effect came tantalizingly close to being significantly replicated, an outcome that encourages further research. Whereas it now appears that the negative test atmosphere was not a crucial factor in causing tight variance with the Dutch sample, a true nonthreat control condition must be introduced before this effect, if real, can be confidently interpreted as a reaction to subconscious threat produced by the subliminal DMT slide. Localization of the tight variance at ±1 deviations is a puzzle for which I can offer no plausible interpretation. On the other hand, it makes sense that the tight variance was prevalent among those subjects who found it easy to take the PET, because this procedure was designed to facilitate psi by reducing conscious cognitive effort.

The two convergent post-hoc correlations involving the imagery items suggest indirectly that when subjects have a spontaneous flash of a target image and act on that flash, the result tends to be a hit. That such spontaneous images might be psi-laden is compatible with what we know about psi in other research contexts. Perhaps the greatest advantage of the PET will prove to be that it provides a mental environment that is conducive to these images in forced-choice ESP testing.

METHODOLOGY

Towards a Reduction of Experimenter Control in the Study of Special Subjects

Dick J. Bierman† *(University of Utrecht, Postbus 80125, 3508 TC Utrecht, The Netherlands) and Johan L. F. Gerding (Parapsychological Institute)*

INTRODUCTION

In a major textbook on parapsychology, a whole chapter is devoted to alternate explanations of ostensible paranormal data (Morris, in Edge et al. *Foundations of Parapsychology,* 70–110, Routledge and Kegan Paul, 1986). The title of the chapter is "What Psi is Not: The Necessity for Experiments," and it is filled with numerous examples of trickery; but nowhere is the link with laboratory experiments explicitly treated. Although the primary purpose of this chapter might be to show skeptics that parapsychologists are no fools, it might also scare away new researchers, since the implicit message is that as experimenters we need more control and should be constantly aware of potential fraud.

Discussions with skeptics, when productive, do result in improvement of controls and procedures, but there seems to be no end to this process. One of us has argued (Bierman, *JSPR,* 1981, 183–184) that whatever the controls imposed, there is an alternative, normal explanation for apparent psi results in any experiment. Even in the retro-PK experiments with a skeptic as an outside observer (Schmidt, Morris, & Rudolph, *JP,* 1986, 1–17), technology and loopholes in the protocol may allow for fraud. An enormous amount of time and energy has been spent to satisfy the needs of the skeptics, while hardly any progress has been made in understanding psi phenomena.

PROOF-ORIENTED VERSUS PROCESS-ORIENTED RESEARCH

Although most researchers pay lip service to the idea that theory-driven, process-oriented research is more important than proof-oriented research, it is obvious from inspection of the above-mentioned and other textbooks that most attention is devoted to proof-oriented research. There is little reference to major theoretical frameworks such as the observational theories, and there is no reference at all to research programs based on a theoretical framework such as the checker effect (Weiner & Zingrone, *JP,* 1989, 203–233). The exception is the theory that internal attention states might affect the detection of psi. With one or two exceptions in fifteen years of research, however, internal attention states have never been formally manipulated.

PROCESS-ORIENTED RESEARCH NEEDS POWER

It is assumed that laboratory experiments in general allow for only small effect sizes. It is argued that the typical single laboratory experiment does not have enough power to establish theory-relevant differential effects, but that meta-analyses can overcome the problem. We think that, apart from using meta-analyses to attain enough power for establishing theory-relevant differential effects, there is another way: using selected, extremely gifted subjects. The use of such subjects, with resulting effect sizes apparently orders of magnitude larger than those found in the laboratory, not only increases power but also might give more insight into the psychodynamics of the paranormal, since no averaging over subjects is required.

SELECTED SUBJECTS AND FRAUD

Parapsychologists are extremely wary of using selected subjects. J. B. Rhine was the first explicitly to abandon research with exceptional subjects, because he had insufficient control. His attitude toward exceptional subjects was reinforced when skeptics, explicitly intending to fool researchers, planted two frauds in the Washington University laboratory.

INVISIBLE MANIPULATION INSTEAD OF TOTAL CONTROL

Instead of forcing indefinite controls upon exceptional subjects so that no alternative explanation for the observed data is possible, we proposed instead to manipulate the experiment in a theory-relevant way and study the data as a function of this manipulation. Obviously, it is more convenient to manipulate only one variable rather than to try to control many unknown factors. Also, by stressing the importance of theory testing the emphasis on fraud prevention lessens, and a better interhuman relationship might develop.

There are two constraints to this approach: (a) the manipulation should be known only to the experimenter and (b) the manipulation should not interact with the possibility of fraud.

The possibility of fulfilling condition (a) is dependent on a theoretical framework. In the context of observational theories, it is often possible to introduce a meaningful manipulation not obvious to the subject. This is also true to some degree for the PMIR model (e.g., manipulation of the system lability). Also, in attempting to fulfill this condition, care must be taken not to change too radically the context in which the subject normally performs.

The effect of the hidden manipulation on fraud (condition b) can only be predicted if all conceivable means of fraud are known. This is of course impossible. Even the use of a magician cannot guarantee the elimination of all potential tricks. Some caution will be necessary when evaluating differential effects, but this problem is largely compensated for since the researcher does not need to worry about excluding all means of fraud.

A CONCRETE EXAMPLE

In the fall of 1991 one of J.G.'s students, A.K., returned the results of two self-administered forced-choice card color guessing tests.

The scores were so high that it was obvious that, if these data were not the result of conscious or unconscious fraud (error could hardly explain them), then we had run into the best-scoring subject ever.

What would have been the standard reaction to these results? Most parapsychologists would have tightened the controls to exclude the possibility of fraud; then the skeptics would arrive and insist on new, even more clever controls. By the time all of this was settled, the phenomena would probably have disappeared (to the satisfaction of the skeptic), and we would have learned nothing. We decided instead to design a procedure that would fit A.K.'s normal procedure of unattended testing and that would give us the potential for scientific discovery.

HIDDEN CONDITIONS

Before the experiment, four new packs of playing cards were randomized for color by an RNG. Three of them were sealed into opaque envelopes and sent to A.K. A fourth envelope was prepared in the same way but filled with a dummy pack of noncolored cards. A.K. was told to guess the colors of the cards in each pack and to enter the guesses on the scoring form. He was also told that he would receive feedback for 3 of the 4 packs, but that for theoretical reasons he would not get feedback for one pack.

The conditions were:

Pack A: correct feedback
Pack B: incorrect feedback
Pack C: no feedback
Pack D (dummy): real pack D stored at a distance; correct feedback

After A.K. filled in the scoring sheet, it was returned to the experimenters, along with the packs still in the sealed envelopes. Pack B was then reshuffled, Packs A, B, and D were sent in an open envelope to A.K., and he was asked to check the packs against his calls.

POTENTIAL OUTCOMES

In the following we discuss potential outcomes of this experiment, with a plus sign indicating significant hitting on a specific pack.

First outcome. A +; B +; C +; D +

This outcome would fit with the traditional (third eye) psi models. A smaller score on D would indicate that it is more difficult to guess a pack at a distance. Fraud is highly improbable, since Pack D is stored safely at a place unknown to the subject. A theoretically meaningful result is obtained.

Second outcome: A +; B -; C -; D +

This outcome would fit with observational models (Millar, *EJP,* 1978, 304–332), which stress the relevance of (correct) feedback. Fraud is improbable, but a theoretically meaningful result is obtained. If scoring on B is positive when compared to the incorrect feedback, this would strongly support the OT's.

Third outcome: A +; B +; C +; D -

This outcome would fit with the traditional local models of psi. Fraud is not excluded.

Any other outcome with one or more of the packs positive is difficult to explain in terms of current psi models but can hardly be attributed to fraud.

DISCUSSION

One could ask whether the advocated loosening of controls and focusing on invisible conditions might be harmful to the field. Current scientific methodology does not adequately deal with n=1 data, yet singular data carry meaning and information. The roots of psi research are in individual cases, and we hope to bring back the study of exceptional subjects as a legitimate way to advance our understanding of psi. The prerequisite is that the researcher must specify a theoretical framework—hardly a drawback.

This approach does not exclude the possibility that skeptics might guess or deduce our hypotheses and try to produce differential effects fraudulently. We believe, however, that as long as the data are presented in a theoretical context, not only as proof of an alternate reality, then the risks are rather small.

Another question concerns any deviation from the original conditions under which the large effect sizes occurred. In our example, the major difference is the delay of feedback. This delay can be reduced by, say, a computerized approach that simulated the original conditions.

The approach suggested in this paper could also be applied to poltergeist cases. Some years ago one of us (Bierman, *EJP,* 1979, 111) proposed filming poltergeist cases with the camera turned on or off, as determined by an RNG and with the conditions unknown to the people involved. Here too, most potential findings would be theoretically interesting.

DISPLACEMENT

Hits on Repeated Targets in Relation to Hitting, Missing, and Displacement

James E. Crandall (Department of Psychology, University of Idaho, Moscow, ID 83843)*

Previous research has shown a fairly reliable tendency for psi-missers in forced-choice ESP experiments to score above chance on targets adjacent to the direct one that is missed. This has been referred to as the psi-missing displacement effect, or PMDE. An unmotivated inhibition of information model appears to be the best way to account for this relation between missing and displacement.

The inhibition model assumes that as a result of some form of cognitive error, the signal from the direct or intended target is sometimes inhibited, leaving the person receptive to signals from adjacent targets. In brief, missing leads to displacement to an above-chance degree.

One implication of this model is that psi-missers should more frequently hit targets that are in the first position of a repeated sequence (doubles, triples, etc.) compared to hits on nonrepeated targets. Receptivity to the signal from the second listing of the target could have a reinforcing effect that would help overcome inhibition to the signal from the direct target.

For brevity, targets in the first position of a repeated sequence

*Presented by Gertrude S. Schmeidler.

will be referred to as First targets and nonrepeated targets will be referred to as Single targets. Later targets in a repeated sequence were not included in the analysis because subjects' frequent aversion to repeating calls could artifactually influence hit rates on such targets.

Based on the unmotivated inhibition of information model, it was hypothesized that missers would show more hits on First than on Single targets. Hits on the two kinds of targets were also examined for hitters. However, this analysis was exploratory and no hypothesis was made.

DESCRIPTION OF STUDIES

The analyses involved 12 forced-choice experiments using five equiprobable targets, with one to five runs of 25 trials. Ten of the experiments were my own; the other two were conducted by Kanthamani (Crandall & Kanthamani, *RIP* 1986, 66–69) and Broughton (*JP*, 1992, 31–38). Each experiment included two conditions, with one being considered to be more psi-conducive than the other. Only data from the favorable conditions were examined here.

RESULTS

Comparison of hits on the two kinds of targets was made by the standard formula for the analysis of differences in proportions, using the correction for continuity where needed.

Missers

In 11 of the 12 experiments hits were greater on First targets than on Single targets. In four of these cases the results were independently significant at $p < .05$. Mean proportions of hits were .1465 for First targets and .1106 for Single targets. The overall significance of the difference in hits on First and Single targets was assessed by Stouffer's z. This involves the sum of the z scores from the 12 experiments, divided by the square root of 12. This yielded $z=3.425$, $p < .0004$, one-tailed. The results supported the hypothesis that later targets in a repeated sequence reinforce the signal from the First target.

Hitters

In 10 of the 12 experiments hits were *lower* for First than for Single targets. Mean proportions of hits were .2610 for First targets and .3009 for Single targets. The Stouffer z for the difference in hits on First and Single targets was −2.688, $p < .007$, two-tailed.

The results from the hitters were somewhat surprising. They showed a reasonably consistent, significant, and opposite pattern compared to the missers. The question is why they should show lower levels of hitting on First than on Single targets. Tart's transtemporal (or transpatial) model provides a possible answer. If, as suggested by Tart, hitters have a tendency to inhibit information from adjacent targets this could weaken, rather than reinforce, the signal from the direct target when it is part of a repeated sequence.

SOAL AND PRATT'S ANALYSIS OF A HITTER'S DATA

After reading an early draft of this paper, Gertrude Schmeidler (personal communication, December 22, 1991) called my attention to an article by Soal and Pratt (*JP,* 1951, 192–215) which involved a reanalysis of Mrs. Stewart's responses in 35,000 ESP trials conducted from 1946 to 1950. Her average scoring rate was well above chance. Similar to the hitters in the present case, Mrs. Stewart made significantly fewer hits on First than on Single targets (χ^2=206.76, 1 df, $p=10^{-44}$). They found that scoring was significantly higher for the First target of a double than for the First target of a longer sequence (χ^2=14.22, 1 df, p=.0002). Soal and Pratt concluded that "ESP is a process occurring with a perspective which is not limited to the individual target" (*JP,* 1951, 209).

Soal and Pratt's findings prompted me to compare hits on First targets of doubles with hits on First targets of longer sequences. For hitters, the two proportions were very similar (.2596 and .2685, respectively). However, the results for missers provided a mirror image of Soal and Pratt's findings with a hitter. The proportions of hits were .1311 for First targets of doubles and a surprisingly high .1953 for First targets of longer sequences ($z=2.098$, $p < .04$, two-tailed). The results suggest the possibility of reinforcing effects coming from more distant targets.

The similarity of the results of Soal and Pratt to the present findings suggests that the effects of target repeats upon hitting may be stable phenomena that occur in many other sets of data. The present results point to the importance of segregating the data according to hitters and missers.

ANALYSES FOR ARTIFACTS

Two further analyses were used to check for the possibility of some artifact influencing the results. The first involved comparing the difference in hits for First and Single targets shown by subjects with very few call repeats to the difference shown by subjects with more call repeats. This was done separately for missers and hitters. No differences were found between the two subject groups.

The second analysis involved scoring each subject's calls against another subject's targets. This control analysis resulted in differences in hits on First and Single targets that were nearly zero for both missers (−.0131) and hitters (−.0038).

In summary, neither the division of the data according to number of call repeats nor the control scoring gave any sign of an artifact that might have influenced the results.

QUANTITATIVE FIELD INVESTIGATIONS

Quantitative Investigation of a "Haunted Castle" in New Jersey

Michaeleen C. Maher[†] *(New School for Social Research, Social Sciences Department, 66 West 12th Street, New York, NY 10011) and George P. Hansen (Cranbury, NJ)*

This paper reports another in a series of experimental field investigations (Maher & Schmeidler, *JASPR,* 1975, 341–352; Maher & Hansen, *JASPR,* 1992, 347–374) that combine the quantitative methods of experimental parapsychology with the use of instruments to detect, measure, and probe the anomalies attending a reported haunting. The authors are empirically-oriented experimental parapsychologists, and thus our approach to the study of survival is "data driven." We wish to understand more than just the reports and the witnesses—we seek to interact directly with the phenomenon itself, if possible. We believe that with the use of mechanical devices, electronic instruments, and human beings as "sensors," we should be in a stronger position to grasp the nature of the reported phenomenon and its implications for the survival of human personality.

CASE

During 11 hours of tape-recorded interviews, 13 witnesses provided accounts of their own and others' experiences in a Norman seventeenth-century-style castle in northern New Jersey. The castle was believed to be haunted, and the disturbing incidents had spanned a period of 50 years. A notion prevailed that the original owner, who had died with "unfinished business" (he had not yet completed building the castle), was caretaking the premises in spirit form. A variety of phenomena suggestive of both haunting and poltergeist occurrences were reported, included several apparent apparitions; inexplicable footsteps; cold spots; paranormal object movements; doors, windows, cupboards, and drawers opening or closing unaccountably; ethereal organ music; sulfurous odors; peculiar malfunctioning of mechanical, electrical, and electronic equipment and devices; and odd behavior by animals.

EXPERIMENTAL TESTS

Sensitives' and Skeptics' Tests

Quantitative tests were administered to six experimental participants using Schmeidler's technique (*JASPR,* 1966, 137–149). Three sensitives toured the castle attempting to sense ghostly presences. They marked the locations of any such perceptions on floor plans. Three skeptics used floor plans to mark sites where, in their judgment, someone credulous might likely have reported ghostly phenomena. The six participants also evaluated a checklist composed of brief descriptions drawn from the witnesses' reports randomly interspersed with plausible but irrelevant descriptions. Sensitives circled items they believed to be consistent with what they had experienced and crossed out items that were not. Skeptics circled items they surmised might resemble prior reports and crossed out items they deemed to be irrelevant. Participants were not required to respond to every checklist item. One sensitive provided description-checklist responses that were unscoreable.

Sensitives' and Skeptics' Test Results

Responses to the floor plan and description checklist measures were evaluated with the Fisher exact method which directly gives a one-tailed probability value. For a discussion of this method, see

Maher & Hansen (*JASPR,* 1992, 347–374). One sensitive reported locations that were in significant agreement with sites where phenomena had previously been reported by witnesses (p = .033). Results for the two other sensitives were at chance (p = .326, p = .394). None of the three skeptics provided floor plan data that significantly matched the witnesses' reports (p = .188, p = .764, p = .649).

Description-checklist tests were significant for one of two sensitives who provided scoreable checklist data (p = .029), but nonsignificant for the other (p = .636). No skeptic's test score on the description-checklist measure was significant (p = .267, p = .914, p = .070), although one score bore a suggestive resemblance to the witnesses' reports.

In order to obtain group scores for sensitives and skeptics, a computer program was written to determine, for each of the four categories (i.e., sensitives on the floor plan, skeptics on the floor plan, sensitives on the checklist, skeptics on the checklist), all possible combinations of possible individual scores that could arise with constant marginal totals of the Fisher exact matrices. This produced four distributions of score sums. Each actual group sum was compared to its appropriate distribution.The sensitives' floor plan score was found to be significant (p = .014), whereas the skeptics' corresponding floor plan score fell within chance limits (p = .371). For the description-checklist test, the group score for the sensitives was also significant (p = .035), whereas that for the skeptics was not (p = .152). (We wish to thank Betty Markwick for suggesting this method and for verifying our computations.)

EXPLORATORY TESTS

Random Number Generator (RNG) Tests

Quantitative tests with an RNG were conducted according to the method used in the authors' prior "haunting" report (see *JASPR,* 1992, 358–359 for a description of the method). Fourteen sites where unusual phenomena had been reported were contrasted with fourteen sites where nothing out of the ordinary had been reported. The length of the test runs varied between 5 and 15 minutes. For each site with more than one run, an unweighted Stouffer Z was calculated. A t-test compared target- and control-site scores ($t = .043$, 26 df, $p =$ n.s.), and an F-test compared target- and control-site variances ($F = 1.304$, 13/13 df, $p =$ n.s.). No significant differences were found.

Photography Tests

No unusual effects that bore further study emerged from Polaroid or infrared photography tests conducted at the castle.

Video Tests

A curious effect was found in a black and white (B&W) video test of an area where phenomena had been reported by several witnesses. The effect was not observed during the videotaping, but appeared on a color monitor when the B&W videotape was being transferred to VHS format. It can be briefly described as a yellow spot (roughly oval in shape) which occupied approximately 5% of the screen area in the upper right portion of the color monitor. A reddish-brown area could be distinguished in the right-hand portion of the oval shape. The spot emerged gradually during the sequence, increased in intensity, and then waxed and waned at irregular intervals. The subsequent sequence did not contain a yellow spot or any other unusual effect. Nor did the prior portion of the reel, or any other reel, contain a conspicuous anomaly. As it is unusual for color to appear in a B&W videotape, even when viewed on a color monitor, an expert from a video transfer firm was consulted. He was unable to explain the cause of the effect and had not encountered any similar effect.

Photo Line-up Test

A 1929 photograph of the original owner of the castle was used to create a photo line-up test to determine whether witnesses would identify the owner's likeness as similar to the apparitions they had reported. Three photographs of males from the same era were obtained from a photo archive. The four pictures were randomly ordered and presented. One percipient chose a picture of someone other than the original owner of the castle as best representing the likeness of the apparition he had reported. Another chose the original owner's picture, but said afterward, when shown a snapshot of the owner with his two sons, that the owner's eldest son looked more like the apparition she had experienced than the owner did. A third, already familiar with the photograph, reiterated his conviction that the owner's picture resembled the apparition he had reported.

Psychological Tests

Projective tests were given to seven witnesses to evaluate their psychological states and dispositions and to see if any witness showed evidence of unusual psi ability. The tests were evaluated "blind" by G. R. Schmeidler, whose work on psi and personality is familiar to parapsychologists (*Parapsychology and Psychology: Matches and Mismatches,* McFarland, 1988). Confidentiality requirements prevent the full disclosure of the test results, but the authors value the insights that such tests can provide. Eventually, when the database is sufficiently large, we hope that the results of such tests can be integrated with the phenomenological characteristics of cases and that a quantitative assessment of the utility of the projective-test profiles can be made.

DISCUSSION

Although no trenchant statements based on the outcome of this research can be made about the general problem of survival, there is much that can be learned from the study of such cases. This case suggests the classic haunting scenario, where a surviving personality with "unfinished business" is presumed to be lingering in a locale and interacting with humans over a period of many years. A variety of witnesses gave credible reports of haunting and poltergeist-type experiences that seemed to cluster in specific locations. The results of the sensitives' and skeptics' quantitative tests buttress the notion that something paranormal is occurring at the site. Yet when three percipients of apparitions were tested to obtain evidence bearing on the identity of the "ghost," the results suggest that witnesses' psyches play an active dynamic role in shaping the perceived characteristics of the phenomena they experience. The study of "ghosts" has faced numerous such difficulties and uncertainties over the years, but we believe that we have developed an approach that may begin to sort fable from fact.

An Investigation of Reported Haunting Occurrences in a Japanese Restaurant in Georgia

William G. Roll†, Michaeleen C. Maher†, and Barclay Brown (W.G.R. & B.B.: Parapsychological Services Institute, Isleway House, Fairfield Plantation, Villa Rica, GA 30180; M.C.M.: New School for Social Research)

Apparent haunting and poltergeist occurrences were reported in a Japanese restaurant in Georgia by the manager, head waitress, cook, bookkeeper, and a visitor to the site. The manager described a recurrent apparition that he had observed. The head waitress and cook described apparitions that were dissimilar to the manager's and to each other's. Cold spots, footsteps, dishes vibrating, electronic equipment failures, and electrical and plumbing malfunctions were reported, as were mysterious nocturnal occurrences such as the deep fryer apparently turning on by itself and a locked fire escape door repeatedly opening and setting off an alarm. Both qualitative and quantitative approaches were used to study the phenomena and to attempt to understand the physical, psychological, and neurophysiological processes that might underlie them.

QUALITATIVE APPROACH

W.G.R. and B.B. recruited Ms. Hayes and three other psychics from the Patricia Hayes School of Inner Sense Development (Hayes also brought five of her students) for a session combining the traditional spiritualist seance and the "empty chair" technique of Gestalt psychology. The purpose of the empty chair procedure in clinical psychology is to allow clients, by verbally addressing their adversaries (imagined as sitting in the empty chair), to experience a meaningful configuration or gestalt and thereby resolve their conflicts. The procedure was adapted to the investigation in the hope that the underlying needs of the witnesses, presumed to be an important aspect of their experiences, would emerge during the session.

The psychics, after touring the restaurant to get impressions of the ghost, met with the manager, the head waitress, W.G.R., and B.B. The participants sat in a circle, and the ghost was invited to join the group. The psychics then volunteered their impressions. They first described a tall, thin, vigilant ghost and then brought in a shorter, older, and heavier ghost with a drinking problem, the manager emphatically agreeing that he had experienced both. (During his initial interview, the manager had described only one recurrently appearing figure.) When W.G.R. and B.B. suggested to him that the two dissimilar apparitions might reflect different aspects of his needs, the manager appeared to acquiesce and began to entertain the notion that the two ghosts were projections of his own personality.

W.G.R. and B.B. felt that the psychics had been successful in enabling the manager to discover a meaningful pattern behind his experiences. The "psi session" and the insights that the manager and investigators derived from it constitute the essence of the qualitative aspect of the study.

QUANTITATIVE APPROACH

M.C.M., who was invited by W.G.R. at this stage to collaborate on the investigation, advocated an experimental approach. As in her prior haunting research, M.C.M. favored using a quantitative field technique originated by Schmeidler (*JASPR,* 1966, 137–149), along with various instruments to attempt to detect unusual physical phenomena. W.G.R. was in accord, and he recruited 10 student volunteers from West Georgia College to act as subjects for the experimental tests. Based on their responses to a psychic experiences questionnaire, W.G.R. classified five students as "sensitives" and five others as "controls." The students were asked to tour the restaurant and to record on floor plans and checklists the locations, personality characteristics, physical characteristics, and activities of the presumed ghost. The students were instructed to request additional checklists if they sensed or inferred the presence of more than one ghost, but none did so. The manager provided separate checklist descriptions for his original and subsequently reported ghosts, but since no student sensed additional ghosts, only the checklists for his original ghost were used for comparison. B.B. provided a target floor plan based on information obtained by the witnesses during interviews.

W.G.R. conducted the experimental tests in M.C.M.'s absence, but unforeseen difficulties arose. Not all students filled out every checklist; nor did all participants respond to every item on the checklists. Moreover, the manager made a sudden appearance at the restaurant and came in contact with several students while the experiment was in progress. (The students did not believe that their test responses had been influenced by the presence of the manager.)

RESULTS OF THE QUANTITATIVE TESTS

The experimental data were sent to M.C.M. in New York City for evaluation. Results for the floor plan test, the "Activity of the Ghost" checklist, and the "Personality of the Ghost" checklist were evaluated with the Fisher-exact test (see Maher & Hansen, *JASPR,* 1992, 347–374, for a discussion of this method which calculates a one-tailed probability value). None of the nine students who provided scoreable floor plan data achieved a significant result. Nor did any of the seven students who provided scoreable "Activity of the Ghost" checklists show a test result that bore a significant positive relationship to what had been reported. One of seven students (a "sensitive") who filled out the "Personality of the Ghost" checklist described an impression that bore a significant ($p=.016$) relationship to the manager's original ghost.

Eight students filled out the "Physical Characteristics of the Ghost" checklist. For each respondent, a weighted score for the completed items, based on the number of possible answers, was calculated. A two-sample t-test between scores for sensitives and those for controls gave $t(diff)=6.505$, 6 df, $p=.001$, two-tailed, showing that the group of sensitives gave descriptions of the ghost's physical characteristics that resembled the manager's description significantly more than did the descriptions provided by the controls. However, because the experimental protocol had been compromised by students' contact with the witness, the test outcomes were regarded as inconclusive.

TESTS WITH PHYSICAL INSTRUMENTS

M.C.M. visited the restaurant to determine whether areas where witnesses had reported ghostly phenomena contained physical anomalies that could be documented by instruments. Using a magnetometer (Alphalab Tri-Field Meter with a magnetic setting measured in milligauss and calibrated for 60 Hz sine waves), she discovered that magnetic fields caused by the heavy industrial equipment and electrical wiring appeared particularly strong in locations that had been implicated in the witnesses' reports. A t-test of the differences in average magnetic field strengths between target locations reported by witnesses as sites of ghostly phenomena and control locations where nothing unusual had been reported gave a significant result (t[diff]=2.957, 23 df, p=.007, two-tailed). Tests with infrared and Polaroid photography did not provide evidence of anomalies.

PSYCHOLOGICAL TESTS

M.C.M. also administered projective psychological tests to the manager and head waitress to learn about their dispositions and psychological states and to see whether either of them evidenced a personality that was consistent with more than average psi ability. The tests were evaluated by G. R. Schmeidler (an expert in the psychology of personality), who was not told any particulars of the case. Neither witness showed evidence of mental derangement, but Schmeidler suggested that PK ability might be consistent with one respondent's (the manager's) personality, while the other (the head waitress) might conceivably become "the victim of a poltergeist but surely not the person causing the poltergeist activity."

NEUROPSYCHOLOGICAL QUESTIONNAIRES

Neuropsychological questionnaires, modified from Neppe (*JASPR,* 1983, 1–29), were administered to the manager and head waitress by W.G.R. The responses were evaluated by M. A. Persinger, a Canadian professor. He reported to W.G.R. that the manager's responses, which described sensations of smells (rotton eggs, sweetness, perfume) and sounds (buzzing, music, words), as well as experiences of dizziness, confusion, and fear when there were no apparent reasons for such experiences, were suggestive of temporal lobe epilepsy. The head waitress's responses were considered to be within normal range.

DISCUSSION

Based on the significant difference in magnetic field strengths between target and control sites, M.C.M. suggested that the strong magnetic fields occurring at sites where haunting phenomena were reported might have caused subtle nervous system changes and even transient hallucinatory episodes in the manager and other percipients. These aberrant sensations (occurring sometimes in conjunction with plumbing or electrical wiring malfunctions that could have been of natural origin) may have led witnesses to wrongly suppose that they were experiencing ghostly phenomena.

Persinger has communicated to W.G.R. that magnetic fields in the 8–10 milligauss range have been known to trigger temporal lobe effects and have been used to generate such effects in the laboratory. M.C.M. observed that magnetic fields at the target sites averaged 8.93 milligauss as compared to 3.08 milligauss at control sites. (Target-site magnetic fields sometimes reached 15 milligauss and several were as strong as 50 milligauss.) The indication of temporal lobe lability in one of two witnesses who filled out W.G.R.'s neuropsychological questionnaires provided further reason for the experimenters to believe that magnetic field strengths at the restaurant played a role in the witnesses' subjective experiences.

W.G.R. suggested that the magnetic fields might have interacted with witnesses' nervous systems to produce the physical (PK) incidents that were reported at the restaurant. M.C.M. cautioned that because it is difficult and expensive to exhaustively rule out natural causes as agents of electrical or plumbing malfunctions, professional experts had not been consulted, and the possibility that these sorts of problems had a natural basis remained. Also, since the deep fryer reportedly turned on and the fire escape door opened during times when no one was known to be at the restaurant, a person-centered PK hypothesis would imply not only that macro-PK occurred at a distance, but that it occurred when the ostensible PK agent was at a distance far removed from the magnetic fields.

Testing the ESP Claims of SORRAT

Richard Wiseman[†]*, John Beloff, and Robert L. Morris (Koestler Chair of Parapsychology, 7 George Square, Edinburgh University, Edinburgh, EH8 9JZ, UK)*

This paper describes a recent assessment of the ESP ability that a well-known parapsychologist (W.E. Cox) attributes to the spirit "entities" which communicate with the Society for Research in Rapport and Telekinesis (SORRAT). The paper first notes that Cox claimed that these "entities" were able to divine the order of a sealed deck of ESP cards. The paper then outlines how an initial pilot study was refused by Cox, in favor of more formal experimentation. The design of this formal experiment is then outlined, emphasizing the controls that were employed to counter potential subject cheating. The experimenters randomized a deck of ESP cards and noted their resulting order.These cards were sealed and dispatched to Cox. The "entities" attempted to divine the order of the sealed deck, and Cox then returned this predicted card order and the sealed deck to the experimenters. The order of the cards was compared to the order predicted by the "entities." The "entities" failed on three counts. First, the deck was not returned to the experimenters by a deadline designated before the experiment. Second, when the deck was eventually returned, the "entities" obtained only 8 hits ($p=0.11$, two-tailed). Third, the "entities" claimed to have paranormally rearranged the order of the cards in the target deck. However, when the deck was opened, all 25 cards were in the same order as when they were dispatched by the experimenters. Finally, the paper discusses the implications of this experiment for future work with SORRAT.

Part 2: Symposium

EUSAPIA PALLADINO AND PSYCHICAL RESEARCH*

Gifted Subjects' Contributions to Parapsychology: The Case of Eusapia Palladino

Carlos S. Alvarado (Institute for Parapsychology, 402 N. Buchanan Blvd., Durham, NC 27701–1728)

A historical view of parapsychology should consider the role of gifted subjects (psychics and mediums) beyond their role as generators of the phenomena that researchers in the field study. Gifted subjects' performances provide a context in or a focus by which the field has developed its methods, theories, and social structure. These subjects are catalysts, focusers of the attention of the parapsychological community.

The Italian medium Eusapia Palladino (1854–1918) was such a catalyst. Her mediumship served functions other than providing evidence for the reality of the so-called physical phenomena of spiritualism. The coverage of her mediumship in the specialized and popular press publicized and popularized physical mediumship and psychical phenomena, thus affecting the field's development. Cesare Lombroso's and Hereward Carrington's

*Chaired by Michaeleen C. Maher

writings about Palladino provide examples. Both of them argued for the legitimacy of the phenomena of psychical research in general, and through their writings Palladino became a publicist for psychical research. That this publicity was not always positive is evident from an analysis of articles and notes published in the *New York Times* about her visit to New York during 1909 and 1910. This analysis shows that the medium's performances generated many controversies and negative images of psychical research that emphasized carnival-like performances, lack of control, and fraud.

Palladino's mediumship also served as a beacon for or catalyst of developments in psychical research regarding theoretical ideas about paranormal concepts of force, considered by many to be the basis of telekinesis, materialization, and other physical phenomena. Her performances (including the phenomena and her behavior) were an important factor in nurturing such theoretical models, because for a long period she was the main physical medium active in Europe. An old tradition of psychic forces found fertile ground in the medium's performances, as seen in the writings of Cesare Lombroso, Oliver Lodge, and Albert de Rochas, among others.

Palladino's mediumship is an example of the influence that gifted subjects can have on the development of psychical research in ways other than providing evidence for the reality of the phenomena under study. It is also thus a cautionary tale concerning the common assumption that parapsychology's history is made only by its researchers. Eventually, perspectives such as this may expand the historiography of this field by enriching our appreciation of the variety of individuals engaged in it and the variety of ways of influencing a discipline's development.

Methodological Recommendations Derived from the "Feilding Report"

Richard Wiseman[†] *and Robert L. Morris (Koestler Chair of Parapsychology, 7 George Square, Edinburgh University, Edinburgh, EH8 9JZ, UK)*

In 1909 the Society for Psychical Research (SPR) published a remarkably detailed account of an investigation into Eusapia Palladino. For over 80 years this document (the "Feilding Report") has been seen by some as containing evidence for directly observable psi. This paper describes how a recent reassessment of the Feilding Report has helped develop methodological recommendations for the assessment of psychic claimants. The paper first briefly reviews the Feilding Report. Next, it outlines a recent reassessment of this remarkable document, which centered around one counterexplanation, namely, that an accomplice may have been present within Palladino's seance cabinet. The paper notes that this hypothesis, if valid, is capable of explaining the majority of the reported phenomena and that the controls against fraud, as described in both the original report and some subsequent literature, are inadequate, as they do not effectively counter this hypothesis. The paper then describes five methodological recommendations which have been derived from this reassessment. The first of these is aimed at researchers engaged in the testing of psychic claimants. The next three recommendations are designed to help researchers report such studies. The last recommendation is for researchers reassessing past studies of psychic claimants. Finally, the paper argues that past studies may prove a valuable resource to researchers interested in developing methodological tools for use in the testing, reporting, and retrospective assessment of psychic claimants.

The Infallible Observer: Cesare Lombroso and Eusapia Palladino

Nancy L. Zingrone (Institute for Parapsychology, 402 N. Buchanan Blvd., Durham, NC 27701–1728)

Some modern scientists appeal to the scientific stature of their forebears as a way of adding to the credibility of the phenomena those predecessors studied. In parapsychology we often use statements about the scientific eminence and perceived skill of the Nineteenth-century Italian psychiatrist Cesare Lombroso, for example, to enhance our claims for the reality of the phenomena provided by the Italian medium, Eusapia Palladino.

There are a number of problems with this rhetorical device. First, little effort is made to determine whether a particular predecessor was indeed eminent in his or her own time or scientifically skilled. Instead we generally repeat a single congratulatory statement from another predecessor to whom eminence has also been attributed. Second, there is no sure relationship between attributions of eminence or scientific skill and specific observational instances in a scientist's working life. Third, and more importantly, even unassailably eminent and methodologically skilled scientists choose their lines of research, develop their instruments, gather their data, and analyze and interpret their results within a social and intellectual context that profoundly influences every aspect of their working lives.

Cesare Lombroso's relationship to Eusapia Palladino provides fertile ground for examining the influence of belief, social theory, and personal relationships on scientific judgment. The following factors shaped Lombroso's belief in survival in general and in Palladino's phenomena in particular: 1) Lombroso's position as a self-educated Jew in a Catholic society; 2) his experiences as a physician with the Italian army in the Southern Provinces during a time of political upheaval; 3) his experience in Psychiatry, Criminology, and Anthropology as empirical delineators of the physiology and physiognomy of the lower social classes; 4) the intellectual climate in which Social Darwinism and degeneration led to an evolutionary justification of both the separation of and the hierarchy among social classes; and 5) the primacy of positivism in the science of the day in which detached and objective observation and measurement were not only possible but an "obvious" consequence of a scientific education.

Although we cannot fault Lombroso for being unaware of the social and intellectual streams that shaped his science, we can draw lessons from his experience by examining his life and work more critically. Lombroso's relationship with Palladino cautions us to become more aware of the possible social and personal forces that shape our own work and, by becoming more aware, to refine our methodology and interpretive schema to lessen the impact of such forces.

Part 3: Panel Discussion

WORKING WITH THE MEDIA: THE SOUND THAT BITES*

Parapsychology and the Media

Kerry Gaynor (2523 23rd St., Santa Monica, CA 90405)

Dealing with the media is a necessary evil that is so heavily laden with pitfalls that one often wonders whether it is worth the effort. But, like complaining about the awkwardness or discomfort of a first date, one is faced with the question, what is the alternative? The bottom line is that we have to get the word out, and the word is that there are serious researchers using scientific method to study such elusive phenomena—indeed, that there are scientists all over the world who have devoted their lives to unraveling some of the most exciting, perplexing, and enigmatic mysteries that have baffled and literally held us spellbound since the beginning of time. This is something to be proud of, and we need to communicate that pride, as well as the progress in our research, to the general public.

This of course brings us back to that necessary evil, the media. The public at large does not even know that the PA exists, but they sure know who the "not so amazing Randi" happens to be. I have worked on many radio shows where people would call in and ask

*Convened and chaired by Michaeleen C. Maher

me what I thought of Randi, instead of inquiring about my previous discussion. In addition, through the years when I have told people that I was going to a parapsychology conference, they have invariably responded, "Oh you mean a psychic fair." This is disturbing, and it demonstrates how poorly we are using the media to educate the public about what parapsychologists really do.

Television obviously has the potential to reach the largest audience, but here the pitfalls are the greatest. First and foremost, the agenda of the show and that of the parapsychologist are often at odds. Of course, that's like saying that the agendas of Dan Quayle and Murphy Brown are sometimes at odds. The distributor wants ratings and to that end he will pursue two things: sensationalism and controversy.

The controversy can be accomplished by setting the parapsychologist up to be ambushed by Randi or one of his cohorts. The sensationalism can be achieved in so many different ways that it's frightening to think about. About a year ago I received a phone call the night before I was scheduled to be on a certain talk show. I was told that I was being cancelled so that they could replace me with a vampire. Producers may also include what you say in the context of someone else's segment, the content of which you might find outrageous and completely lacking in credibility. In addition, if the program is taped, they can edit you, and they always do. You simply have no control over the final outcome. Finally, producers frequently lie to you.

Another problem one is likely to encounter is that the producer of the segment will often know absolutely nothing about the field of parapsychology. He will tell you that he has just completed a story on the language capabilities of an orangutan, and you will soon find yourself wondering whether he was actually the producer or the subject himself. Producers ask me to take them to a haunted house so that they can film a ghost. I tell them that it doesn't work that way: I have spent over eight thousand hours inside these alleged haunts, and most of my time has been spent developing my patience. Their response (ironically enough, the only predictable aspect of my research) is always, "Well, I understand that, but we still need to film a ghost; after all, we can't just have talking heads."

Although there are countless headaches and problems inherent in working with the media, only a few of which have been detailed here, we still must do it. In a lecture we can reach perhaps five hundred people, but in a television show we can reach anywhere from five to thirty million. We must use the media wisely to share what we've learned and to get the word out that we are trying to understand mysterious phenomena, which constitute a major chunk of human experience. The "skeptics" have nothing to offer this society in terms of the advancement of our understanding; they want to just throw it all out. At least parapsychologists offer the hope that someday our knowledge will surpass our ignorance and fear.

Working with the Media: The Sound that Bites!

Michaeleen C. Maher (New School for Social Research, Social Sciences Department, 66 West 12th Street, New York, NY 10011)

My media experience, principally with network TV programs, has taught me the following basic coping strategies.

HAVE AN AGENDA

We have, in fact, more power to influence the final outcome of a TV show or "segment" than is generally appreciated. The producers, when they initially make contact, usually haven't yet fashioned, much less finalized, the TV program that will air a short time later. They anxiously hope to learn from us some tangible facts around which they can quickly structure a suitable story.

It makes sense to carefully think through how best to portray one's work on TV, and to be ready to present that agenda when the opportunity strikes. If our ideas and images are workable, the producers will cheerfully adopt them—and take credit for them too! We may be dismayed to see some of our best material "wind up on the cutting room floor," but the final program will be better off for having considered the topic from our perspective.

When imagining how best to present your research within a narrow time frame, think visually, for TV is a visual medium. Experimental research, particularly, is complex and difficult to portray. Directors rely on the use of "talking heads" only when they can't think of photogenic action shots to move the story along. Think concretely. Are there important facts or caveats that a lay audience needs to be aware of? Can you envision ways to get these points across? Use your creativity to devise succinct visual metaphors that can instruct and enlighten.

Consider sending a brief written treatment to the producer outlining the specific shots that you think would be effective. Tangible descriptions, in the early stages of planning, can influence how your work is approached and how it is ultimately presented.

THE BIGGER THEY ARE, THE HARDER THEY ARE

In my experience, the bigger the budget, the less likely you are to achieve aesthetic or personal satisfaction from the production. Corners can be cut, and they will be. Don't expect to be paid a reasonable sum for your time and effort. If the producers can pass some of the production costs along to you (e.g., your telephone, supplies, and other expenses), they will shamelessly do so. Protect yourself with a signed "letter of intent" that specifically mentions the producer's liability for such costs before you incur them. Request a negotiated release that is tailored to your needs rather than signing a standard one.

THE SOUND THAT BITES!

Communication on TV consists of "sound bites." To be effective, complex ideas must be presented in a disarmingly simple way. Only small chunks of information that drive home points are likely to survive the final edit. Practice sound bite strategies by describing your research to a child under ten. Metaphors may help. Polish your technique, for incoherence can, and will, be used against you. Don't be afraid to discuss your own ideas and theories, as such. In short, to be effective on TV, one must be lucid, lively, courageous, and controversial!

The Parapsychological Association: Working with the Media

Sandra Martin (Paraview, Ansonia Station, Box 929, New York, NY 10023)

As a literary agent I place my clients on documentary, reenactment, and interview television programs. These are some of the guidelines I've learned.

When a television producer calls you, he is probably a segment producer, five steps down from the creator of the program, and knows virtually nothing about what your specialty is; he looks to you to tell him. He has been given your name as someone they "might" want to use. We in parapsychology seem to think that because they call us they must have a larger picture and know what they want. That is usually wrong.

To start off, ask lots of questions, such as: "Is this a pilot, a funded program, or a series in production?"; "Who are the producers and what have they produced?"; "Is this a preliminary interview and are you talking with others of similar background?"; "Who else is involved?"; "What is the overview of the program?"; "How do you plan to include my work?"; "How did you get my name?" Questions such as these will give you the basis for commitment or noncommitment to the program.

Decide exactly what you want to convey and how you want to convey it, within the context of the answers to the questions above. If you want more time, ask whether you can call them back after you've had time to think. When you do talk with them, be concise, simple (the simpler the better for the networks), straightforward, and entertaining (if possible). Try to talk to the director to see whether he/she has a script. If they want names, places, suggestions—in other words if they want you to do their work for them—ask if they have a consulting budget. Every situation is different, and sometimes it just feels good to help focus the program.

The best way to protect yourself is to be clear about what you want to say and say it in as many different ways as possible. On experiential shots you just have to make the best of an awkward situation—unless you can get the director to let you be the "expert" with the overview.

Naturally, you won't be there when the piece is edited, and they will probably take your words out of context, make you look foolish, and ultimately make you wish you'd never agreed to your five minutes of fame.

Working with the Media

Janet Lee Mitchell, Ph.D.
P.O. Box 3115, Bullhead City, AZ 86403

I have been involved with all phases of media, from film documentaries to publishing, since the late 1960s. I've participated in a number of controversial television and radio talk shows, such as Joe Franklin, Barry Farber, and Gary Null. I prefer radio to TV because you usually have an hour or so to share information, and often you get to comment on direct questions from your audience. It bothers me that moderators usually consider having me on their program to discuss ESP only if they can also have a "psychic" or "skeptic" on at the same time. Often the psychic has no interaction with the parapsychological community, or the skeptic has not done his homework. If the person has no knowledge of our methodology or experiments, the so-called controversy is less than intellectually stimulating.

For instance, recently I was pitted against the President of a skeptics group at Arizona State University on a call-in radio show hosted by Tom Leykis. I sent my book *Out-of-Body Experiences* to them, so that they would have a clear idea of what I would be talking about on the show. Neither of them scanned the book, and they seemed to presume that I was someone who went out of my body and taught others to do the same for a fee. They were at a definite loss in dealing with the scientific data I presented on the show. The debate had been scheduled for two hours, but Leykis stopped the segment after an hour to prevent further embarrassment to the skeptic and because the call-in audience was showing interest in the information I shared on OBEs. The preconceived plan for the show had obviously been to ridicule the "out-of-body lady" and have a few laughs at her expense. My preconceived plan for any show is to have three or four basic ideas I want to get across to the audience. Regardless of how asinine or demeaning the process becomes, I keep paraphrasing and reinforcing these ideas in a positive way.

I think publishers have dramatically increased sales by using "new age" rather than "occult" as a publishing category. The word "occult" has negative connotations, and people confusing it with the word "cult" often steer clear of this section in bookstores. The new age category covers all sorts of marginal books from metaphysics to pop psych to holistic health to radical scientific theory. It stimulates ideas of human potential and alternative approaches, and people are much more likely to check out this section, so we have a better market for our ideas under this heading.

Editing can be a horrendous experience, if your editor is unfamiliar with your work and is focused on crass commercialism. For example, I wrote the following for a proposed article: "Latest out-of-body research shows that we can operate independently from our bodies and suggests that we survive death as nonphysical beings." It was edited to read: "Latest out-of-body research proves there is life after death, says a recognized parapsychology expert." I did not allow this to be printed; I also didn't sell the article. I have found that if one's answers are cautious, the audience may consider that nothing is being said. However, if one gives a direct statement (as above), editors will often change it to be more sensational and less factual.

It has been difficult for me, as an academic parapsychologist, to maintain my dignity and integrity in working with the media for over 20 years. My greatest coup was on the Barry Farber radio talk show in New York City, where I was pitted against the Science Editor of the *New York Times* to discuss the possibility of OBEs. He made light of the subject for the first hour or so while the moderator kept calling me "Jeannette, the psychic." It was gratifying in the second hour to be taken seriously, to be called Dr. Mitchell, and most of all, to hear the Science Editor describe his own OBE at age 16!

Parapsychology and the Media

Dean I. Radin (Institute for Advanced Study at Waltham, 253 Bishops Forest Drive, Waltham, MA 02254)

Over the past six or seven years, I have had the opportunity to represent parapsychology as a reputable, interesting, and scientific endeavor to several types of broadcast and print media and to technical, scientific, and popular audiences. I have learned a few lessons.

Although most popular media invariably distort what we hope to convey, the fact that a member of the PA appears on television, as opposed to an overly credulous enthusiast or a debunker, is good. By portraying ourselves as we are—careful investigators and scholars—we can help overcome the silly stereotype of what "parapsychology" has come to mean in the popular mind. In most instances, we cannot control the editorial slant that the media takes, so it is not worth worrying about that. However, we are the experts, so the first lesson (for interviews) is: be prepared. Take the initiative by giving the interviewers or reporters a list of good questions to ask you. Typically, they are pleased to accept your questions, because they are often on deadlines and cannot spend much time on advance research. Don't assume that they know what is interesting or important to ask, because usually they have no idea.

The second lesson is that most popular media, especially TV, are interested in entertainment, excitement, controversy, and scientific "proof." In contrast, most serious science media try to avoid excitement, controversy, and the naive notion of proof. As a result, one cannot simultaneously please the popular and serious media. The question is, if we had to concentrate on one or the other, which is likely better to advance parapsychology as a science? On one hand, the serious media are better because the popular media (and many funding sources) use them as the arbiter of truth. Unfortunately, while the serious media have about the same level of understanding of parapsychology as the popular media, they hold deeper prejudices. So, although the popular media may be better because they are much easier to approach, ultimately I'd suggest working with the media you feel most comfortable with. Personally I find the popular media much more fun, but the serious media more (professionally) profitable.

The third lesson is that the popular media outlets are incredibly powerful in shaping public and scientific opinion on controversial topics. When I give talks on parapsychology to scientific audiences, I usually find that although the talks may attract large audiences, the vast majority of the audience knows virtually nothing about the field. By asking the audience a few questions, one quickly finds that scientists and lay people learn about parapsychology from what they see on TV shows and read in newspapers. What is gleaned from these media is often horribly distorted or dead wrong. So the lesson is as follows (as suggested by Gore Vidal): "Never miss a chance to have sex or appear on television." To which I would add: "or appear in a newspaper or a magazine." In other words, take every opportunity to "popularize" the field, otherwise the only messages the public will see will be wildly sensationalized or equally wildly debunked.

The Sound that Bites: Researchers and the Media

William G. Roll (Psychical Research Foundation, Isleway House, Fairfield Plantation, Villa Rica, GA 30180)

The scientific establishment and the general public are divided about psi. The former is usually skeptical, whereas interest among the latter is increasing. The reason seems obvious. The findings of psi research validate the experience of ordinary people but entail a paradigmatic shift that could shake the establishment.

Increased interest in psi among the public has led to increased interest by the media. This has positive and negative effects. (I limit myself here to field investigations.) Three positive effects: (1) the media may alert researchers to new cases; (2) the media, such as TV producers, may finance investigations that would otherwise not be possible; and (3) the media may enable researchers to reach the public and inform people about psi.

Three negative effects: (1) the need for reporters to get a story may interfere with the conduct of an investigation; (2) the media are likely to present popular interpretations of the phenomena, for instance, in terms of ghosts and poltergeists, and ignore the scientific views; and (3) pressure by the media to get an exciting story may cause some researchers to claim more than they should and cause others to clam up.

Ways to minimize the negative effects: (1) insist on postponing interviews until the completion of the study and then hold a news conference to give all reporters equal opportunities to learn the facts; (2) be prepared to show that the scientific interpretation is more exciting than the popular one; and (3) learn to describe your work in language people can understand—if you use sound bites, be sure they are not the kind that can come back and bite you in the end.

Part 4: Poster Session

Parapsychology Periodicals: Conceptual and Historical Perspectives

Carlos S. Alvarado (Institute for Parapsychology, 402 N. Buchanan Blvd., Durham, NC 27701–1728)

Scientific periodicals are generally considered to be repositories of information and means of communication between scientists, but they are also active agents or vehicles of change and of propaganda, since they express particular ideological, methodological, and theoretical perspectives. Such is the case with parapsychology journals.

In England John Elliotson used the *Zoist* as an instrument for pro-mesmeric propaganda and for defending mesmerism from its critics, as seen, for example, in his frequent criticisms in the *Zoist* of the medical profession's opposition to mesmerism and its phenomena.

Other periodicals, such as Allan Kardec's *Revue Spirite*, were used to publicize and spread spiritism and to defend a particular interpretation of psychic phenomena (the action of deceased individuals).

One of the most influential periodicals in the history of the field has been the *Proceedings* of the Society for Psychical Research. The early SPR researchers used this periodical to give order to the study of psychic phenomena by defining, classifying, and identifying the research problems of the times in its pages. The *PSPR*'s content represents a reorganization of psychical research along the empirical lines that characterized nineteenth-century science in general. Part of this was accomplished by emphasizing methodological prescriptions and evidential problems in the study of apparitional cases and procedures to conduct thought-transference experiments.

Other European journals such as the *Annales des Sciences Psychiques* and its successor, the *Revue Metapsychique,* fulfilled functions similar to those of the *PSPR.*

In the United States, the *Journal* of the American Society for Psychical Research served mainly the interests of James Hyslop during his years as editor. Not only did Hyslop actively use the *Journal* to promote his views about survival and other topics in long articles and book reviews, but 67% of the articles published during that period were authored by him.

Similarly, the *Journal of Parapsychology* was one of the vehicles by which J.B. Rhine articulated his experimental research program. Through the *JP,* Rhine publicized his approach and hoped to change the field's general outlook. His writings in the *JP* were not limited to the experimental approach but extended to his general philosophy about a variety of topics, including professional issues such as terminology, spontaneous cases, Kirlian photography, and experimenter fraud.

The fact that almost every group publishes its own journal or magazine is an indication of the power of the printed word. An awareness of the use of these periodicals alerts us to the influence of hidden and overt agendas in the shaping of knowledge in parapsychology.

Which Half of the Brain Believes in ESP? Testing a Lateralized Version of the MAT (*M*eaning *A*ttribution *T*ask)

Peter Brugger (University of California San Diego, Department of Veterans Affairs Medical Center, Psychology Service [116B], 3350 La Jolla Village Drive, San Diego, CA 92161), Theodor Landis, Marianne Regard, Konrad Mebert, and Barbara Weber (University Hospital, Zurich, Switzerland)

In the Meaning Attribution Task (MAT) subjects rate artificial coincidences (pairs of drawings) according to their meaningfulness. We have shown belief in ESP to be associated with a high MAT rating (*RIP* 1991, 94–98). In the present study we have developed a lateralized version of the MAT, expecting (1) a higher MAT rating for the right cerebral hemisphere (RH) than for the left (LH); (2) a higher MAT rating for sheep than for goats; and (3) a positive correlation of belief in ESP and the amount of RH bias for the lateralized MAT.

Twenty-two right-handed men volunteered in a tachistoscopic experiment which was introduced to them as "an examination of people's ratings of semantic associations." Each MAT pair was presented for 50 milliseconds either to the right visual field (RVF) or to the left visual field (LVF). After each stimulation the subject had to rate the semantic relation of the two drawings by saying a number between 1 ("wide or no relation") and 6 ("very close relation"). The term "semantic relation" was not specified. Immediately after the 20 presentations, subjects had to rate their belief in ESP on a 6-point scale (1, 2, 3=goats; 4, 5, 6=sheep).

MAT ratings were not different for the two visual fields/hemispheres (LVF/RH: mean=2.80, sd=0.69; RVF/LH: mean=2.83, sd=0.58; t=0.77, p > .1). The most plausible reason for this failure to replicate previous findings of a RH dominance for meaning attribution (*RIP* 1991, 96–97) may be that the stimuli were highly structured and thus easy to verbalize.

Over both visual fields, sheep (n=14; mean rating=3.07, sd=0.58) gave a higher MAT rating than goats (n=8; mean rating=2.38, sd=0.45; t=2.93, $p < .005$). The correlation between degree of belief in ESP and mean MAT ratings was found to be significant (Spearman rho=+.66, $p<.001$). Being a replication of our previous finding (*RIP* 1991, 96), this result deserves the same interpretation; belief in ESP may arise from a relative overestimation of the "meaningfulness" of real life coincidences.

The correlation between degree of belief in ESP and the individual MAT score laterality index

$$\frac{\text{mean rating LVF - mean rating RVF}}{\text{mean rating LVF + mean rating RVF}}$$

is significant with $p < .02$ (Spearman rho=+.54). The more pronounced a person's belief in ESP, the higher the MAT rating after LVF/RH stimulation is in relation to that after RVF/LH stimulation. This result not only corroborates earlier findings of sheep's more "right hemisphere style" of thinking but is also in line with recent developments in neuropsychology: Personal relevance as an independent attribute of any stimulus has convincingly been shown to be established, maintained, and processed mainly by the RH. There may therefore be a RH bias in the judgment of the meaningfulness of real life coincidences and thus a stronger inclination to take these as evidence for ESP by a person with a relatively dominant RH.

Correlation Waves, Cognitive Style, and Target Characteristics II: Data from Additional Unselected Subjects

Bruce E. McDonough, Norman S. Don, and Charles A. Warren (Kairos Foundation and University of Illinois at Chicago). Address correspondence to Bruce E. McDonough, Ph.D., Kairos Foundation, 405 N. Wabash Avenue, Suite 2201, Chicago, IL 60611

INTRODUCTION

Don, McDonough, and Warren (*RIP* 1990, 83–87), upon examining five data sets from selected subjects in forced-choice psi experiments, observed that the correlation between calls and focal and displaced targets exhibited wave-like properties. Unlike most previous studies of psi displacement effects, which were generally limited in scope to ± 1, or more rarely, ± 2 displacements, Don et al. examined all twenty-five possible displacement lags in the circular convolution of calls and targets. A spectral analysis of the circular convolution in these selected subjects indicated significant spectral power when compared against Monte Carlo sampling distributions of random data. Furthermore, Don et al. extended these findings to data from a group of unselected subjects who participated in a forced-choice psi experiment conducted by Crandall (*JP*, 1987, 191–215) designed to study the effects of visualizer/verbalizer cognitive styles and high-imagery/low-imagery targets on ± 1 displacements. In the data from unselected subjects, Don et al. sought to determine whether cognitive style and target imagery, shown by Crandall to influence ± 1 displacements, might also effect the power of the correlation wave extending throughout the entire range of displacements. The results of that analysis were suggestive of the presence of a correlation wave in the displacement data from at least some of Crandall's unselected subjects but were inconclusive with respect to the effects of cognitive style and target imagery. That is, post hoc Monte Carlo tests indicated that the spectral power of the correlation wave was significantly greater than chance for a subset of the unselected subjects (visualizers guessing high-imagery targets), but the results of an ANOVA comparing the spectral power from visualizers and verbalizers guessing high- and low-imagery targets, although showing trends in the expected direction, were not statistically significant. As demonstrated by Utts, however, the failure to achieve statistical significance does not necessarily mean that a real effect was not present; there might simply have been too little statistical power to detect an effect below a particular size (*JP,* 1988, 305–320). Because of the significant Monte Carlo results and because the ANOVA trends, though nonsignificant, were in accord with our predictions and made sense from a theoretical point of view, we suspected this to be the case.

We therefore decided to examine additional data to provide for a more powerful test of the hypotheses. Fortunately, such data were already available for analysis. Crandall's 1987 paper reported results from two studies conducted under nearly identical conditions. The data from 59 unselected subjects who participated in Crandall's "Study 2," and which were used in our previous analysis, were combined with data from another 39 subjects who participated in Crandall's "Study 1," yielding a combined total of 91 subjects for the present analysis. We tested the same hypotheses as before, i.e., we pitted the "right hemisphere" hypothesis discussed by Broughton (*RIP,* 1975, 98–102) against the competing "congruence" hypothesis advanced by Maher and Schmeidler (*JASPR* 1977, 261–271).

METHODS

Analyzed for the present study were data from 91 unselected subjects participating in Crandall's 1987 "Study 1" and "Study 2." The task involved clairvoyance; lists of 25 (five sets of five) high- and low-imagery words were used as targets. The subjects were classified with the Paivio imagery questionnaire as having either visual (43 subjects) or verbal (48 subjects) cognitive styles. Subjects were also classified as psi-missers (52 subjects), i.e., those who scored less than 5 hits on focal targets, or as psi-hitters (39 subjects), i.e., those who scored more than 5 hits on focal targets.

The present analyses used circular-match convolution, an analogue of circular convolution. For each of 182 runs, the circular convolution was computed by first reversing the order of data in the target sequence and then counting the number of symbol matches at each of 25 lags. The mean of the components of this circular convolution was subtracted from each point, and the ends were tapered with a Hanning window and padded with zeroes, before power-spectral densities were computed. The maximum power value from the resulting spectrum for each data set (excluding the first two spectral lines) was recorded.

RESULTS

Maximum spectral power was analyzed with a three-factor (Cognitive Style x Focal Score x Target Imagery) ANOVA with repeated measures across the Target Imagery factor. (It is noted that, although the distributions of power maxima were all somewhat positively skewed, F tests are known to be relatively unaffected by violations of the assumption of normality when all populations of interest have the same nonnormal form. Moreover, additional ANOVAs [not reported here] performed after normalization of the data by log-transform yielded comparable results.) Presently, a marginally significant interaction between Cognitive Style (visualizer/verbalizer) and Target Imagery (high-imagery/low-imagery targets) was observed ($F [1/87] = 3.66$, $p = .059$). None of the main effects were significant, nor were any of the interaction effects involving the Focal Score (psi-missers/hitters) factor. A test of the simple effects of Target Imagery in visualizers indicated that the mean power of the correlation wave was greater for high-imagery targets (0.329) than for low-imagery targets (0.261) ($F [1/41] = 4.98$, $p = .031$). A test of the simple effects of Target Imagery in verbalizers indicated that the mean power of the correlation wave for high-imagery targets (0.287) and low-imagery targets (0.310) did not differ significantly ($F [1/46] = 0.34$, $p = .564$).

DISCUSSION

The results of the present analyses, which pooled data from both Crandall's "Study 1" and "Study 2," confirmed our suspicions that a more powerful statistical test would detect the effects of cognitive style and target imagery on the correlation wave in unselected subjects. Crandall found a pattern of ± 1 displacements more consistent with the congruence hypothesis advanced by Maher and Schmeidler than with the right hemisphere hypothesis advanced by Broughton. In contrast, the present analyses of spectral power extending throughout the entire range of displacements were more supportive of the right hemisphere hypothesis than the congruence hypothesis. The presently observed interaction between cognitive style and target imagery was consistent with both hypotheses; predictions derived from the competing hypotheses differed only in the patterning of the interaction. The right hemisphere hypothesis led to a prediction of greater power in the correlation wave for high-imagery targets than for low-imagery targets in visualizers, and no differences between high- and low-imagery targets in verbalizers. This is exactly what we found. On the other hand, the congruence hypothesis led to a prediction of a crossover interaction. In addition to greater power for high-imagery targets than for low-imagery targets in visualizers, it predicted a complementary advantage for low-imagery targets over high-imagery targets in verbalizers. This latter prediction was not confirmed; however, absolute differences in the predicted direction, although nonsignificant, lent some minimal support also to the congruence hypothesis.

There is a consistency of pattern between these results and the spectral analyses of data from selected subjects reported by Don et al. Visualization is the common denominator. Although in our study of selected subjects there was no testing for cognitive style, visualization techniques were used by the selected subjects in those runs for which significant correlation waves were observed. The Zener symbols used in our studies of selected subjects were, analogous to Crandall's high-imagery targets, also highly visualizable. Also, Don et al.'s effects were not restricted to psi-missers (although Crandall's were). In the present study, both psi-hitters and psi-missers contributed to the observed effects.

Comparison of the results between the present study and Crandall's 1987 experiment may also be instructive. As noted above, Crandall's analysis of localized (± 1) displacement scores supported the congruence hypothesis, whereas the present analyses of the correlation wave were more supportive of the competing right hemisphere hypothesis. This pattern of results might suggest that the global, holistic character of the correlation wave is more amenable to right hemisphere cognition, which is visuospatial and holistic, in contrast to left hemisphere cognition, which is verbal and analytic. Left hemisphere cognition involves sequential processing of discrete information, whereas the right hemisphere performs simultaneous processing of diffuse information. Therefore, visualizers may cognize all similar targets, thereby producing nonlocalized displacement effects, i.e., correlation waves. Verbalizers may focus on intended targets, thereby producing only localized (i.e., ± 1), displacement effects.

In the physical sciences, the presence of wave patterns is usually accepted as evidence that a physical process is operative. In the present case, the results may be taken as evidence that a lawful, natural (but not strictly physical) process is involved with psi.

Part 5: Invited Addresses

The Lessons of History

John Beloff (Department of Psychology, University of Edinburgh, 7 George Square, Edinburgh EH8 9JZ, Scotland UK)

Some years back I was approached by a London publisher to write a brief history of parapsychology, and in the spring of this year the work was completed and now awaits publication. In it I have attempted to trace the evolution of our field from the time of the mesmerists, or even earlier, down to the present. I always knew it would not be an easy task, but the difficulty was not just the practical one of trying to condense an enormous number of facts into a manageable and readable narrative; the real challenge was, rather, how to sustain the reader's interest in a history that could not, alas, be presented as a straightforward success story—which is what one would tend to do if one were writing the history of any of the conventional sciences.

In saying this, however, I do not want to belittle the efforts of the researchers themselves. On the contrary, their dedication to their science in the face of every kind of obstacle and discouragement can only be described as heroic. But, in relation to the three main criteria of success in conventional science—namely, (1) theoretical insights, (2) practical applications, and (3) intellectual prestige—one cannot pretend that parapsychology (or psychical research—I shall use the terms interchangeably) has yet proved itself. So perhaps we should take as our first lesson of history this lesson in humility. The hard truth we have to face is that not even the bare *reality* of psi phenomena has yet been established to general satisfaction.

THE RELEVANCE OF HISTORY

To be fair to ourselves, however, it must be said that it is only in the physical or natural sciences that spectacular progress has become the rule. We have only to think of what has happened in our lifetime in such fields as, say, medical research, microbiology, information technology, particle physics, or cosmology, to name but a few salient examples, to appreciate what such progress can mean. In the case of the human or social sciences, on the other hand, be it economics, sociology, anthropology, or especially, psychology, what we find are not so much revolutionary discoveries or permanent advances as a succession of shifting viewpoints, or schools of thought, coupled with a progressive sophistication in techniques and methodology. For example, the current use of computer analyses in the social sciences has added considerably to their power. Parapsychology compares not unfavorably with these other human or social sciences when it comes to statistical refinement or innovative instrumentation.

That being said, however, the vicissitudes of parapsychology are much more dramatic than anything we find in such other sciences, especially in what concerns the "strong phenomena." In discussing the history of parapsychology, one needs to distinguish between phenomena which, at face value are self-evidently paranormal (i.e., the strong phenomena) and those phenomena whose paranormality can only be gauged by dint of statistical analysis (i.e., the weak phenomena). The latter, though known since the late nineteenth century, do not become important before the "Rhine Revolution" (as I like to call it). So far as the history of the strong phenomena is concerned, what we find is something much more reminiscent of the history of art than the history of science—be it the natural or the social sciences. For, in the history of art—and this is true of literature and music as well as the visual arts—there is no such thing as systematic progress. What you find instead is the brief flowering of some new school or movement, usually centered around certain individuals of genius, followed in due course by inferior, uninspired, and imitative work. Eventually it dissolves or is eclipsed, after which some new school or movement arises elsewhere, dedicated to quite different ideals, until it too, in turn, becomes exhausted and is superseded. The whole process is so familiar to us that examples are scarcely needed. You have only to consider how hard put one would be to name even one distinguished painter or sculptor in France at the present time!

In like manner, we find in the history of parapsychology, numerous examples of promising new developments, usually centered on some individual or individuals credited with outstanding new powers, which, for a time, make it look as if a new era is, indeed, imminent. But what happens? The phenomena decline, the individuals lose their powers or become discredited, and what started with the highest hopes turns out to be a false dawn. Perhaps the most recent episode to conform to this pattern is the spate of ostensible paranormal metal-bending that followed the advent of Uri Geller in the early 1970s.

Of course, one critical difference between the history of art and the history of parapsychology is that works of art normally remain extant so that one does not have to defer to the historian's judgment. In the case of parapsychology, on the other hand, the phenomena in question are no longer there to be observed so that we are entirely dependent on records and reports. You could say, of course, that this is no different from, say, political history, where we are likewise dealing with bygone events to which only documents or artifacts still bear witness. Paranormal claims are open to suspicion in a way that ordinary historical events are not. Perhaps if the great exponents of the strong phenomena in the past had devoted themselves to producing paranormal artifacts (i.e., permanent paranormal objects) instead of transient materializations or table levitations, we would now be much closer to the position of the art historian—but such was not to be. The point I want to stress, however, is that, like those outbursts of creativity that we associate with the arts, the onset of paranormal phenomena appears to proceed in waves. Why this should be so we shall consider in due course. Meanwhile, I am proposing it as our second lesson of history.

One consequence of these fluctuations is that many strong phenomena, just because they are no longer observable today, tend to be dismissed as spurious even by some who are very knowledgeable about the historical facts. Moreover, not only do the critics of parapsychology, like Ray Hyman or Antony Flew, question the relevance of the historical evidence, but many practicing experimental parapsychologists concur. The position they adopt is that if a given phenomenon cannot now be elicited for experimental purposes, it would be wiser simply to ignore it. Parapsychology, according to this school of thought, should concern itself exclusively with problems which we can hope to resolve. I for one, however, must dissent. If we aspire to become something more than just technicians, we need to preserve a perspective. To confine one's attention to the here and now is, I submit, a mark of provincialism. I like to think of the study of the paranormal as displaying a vast landscape replete with all kinds of weird and exotic flora and fauna, so to speak, which, even if they are currently inaccessible, are too intriguing to ignore. In short, writing a history of parapsychology has convinced me that history does have lessons, that the past is not irrelevant to current problems and concerns.

THE DECLINE EFFECT

One aspect of the uneven course which our science has taken has been this steady attenuation in the strength of the phenomena which we try to study. Ian Stevenson, in his presidential address to the SPR in 1989, spoke of "The Decline of Major Paranormal Phenomena" (Stevenson, 1990). This can be seen most conspicuously in the case of the paraphysical phenomena. Physical mediumship is now virtually a thing of the past, so that the only examples of macro-PK now available for study today are poltergeist cases (Recurrent Spontaneous PK, to use the technical jargon). Interestingly, such cases constitute an exception to the rule inasmuch as they have changed very little down the centuries and show no signs of disappearing. Unfortunately, they are shortlived episodes and, though certainly worth studying, present formidable practical difficulties for systematic or definitive research.

Be that as it may, if you want to study PK, the approved method is to get your subject to try influencing a binary electronic random event generator. Indeed, so much has this become the recognized mode that some contemporary exponents of the technique have cast doubt on whether macro-PK has ever been convincingly demonstrated. At all events, in this computer age, we are now free to run millions of trials in a brief span of time so that even a quite minute PK effect—and Robert Jahn, for example, at his laboratory in Princeton, is happy with a bonus of even one extra hit per thousand trials with an average subject—could still produce an impressive overall score provided, always, that the direction or scoring remains consistent. Unfortunately, very few subjects are capable of such consistency for long enough to produce scores that are any more significant than those that were forthcoming in the bad old days when we had to resort to the laborious process of tossing dice. Indeed, a recent meta-analysis carried out by Radin and Ferrari (1991) shows that the overall "effect size" of the dice experiments is slightly superior to that of the RNG experiments—though both are extremely marginal.

Over ten years ago Edwin May of the Stanford Research Institute obtained funding to carry out what he conceived to be the methodologically perfect PK experiment, using all the devices that modern technology could provide (May, Humphrey, & Hubbard, 1980). His project continued for about one year. At that point, the overall odds against chance amounted to a modest 50 to 1. This is better than a null result, but the thought of so much effort and expense being harnessed to produce such an exiguous outcome must make one worry as to whether we may one day reach a position when it would become virtually impossible to demonstrate PK at all.

The short-term decline effect was, of course, a familiar feature of the parapsychology laboratory when J.B. Rhine was in his heyday. Indeed, the intrasessional declines provided perhaps the clearest clue in the early PK work that something out of the ordinary was going on. Unfortunately, it soon transpired that a decline effect, for ESP no less than for PK, could persist across sessions and, ultimately, across an entire career. Nearly all the high-scorers eventually lost their ability. Even Pavel Stepanek, whose ten-year career as an ESP subject earned him a mention in the *Guinness Book of Records,* eventually ran out of steam. When, after a long break, he was retested recently by Dr. Kappers in Amsterdam, he could produce only chance scores (Kappers et al., 1990). I do not think it was loss of motivation or boredom in his case, as has sometimes been put forward as an explanation for the long-term decline effect. It was Stepanek's great strength that he was constitutionally incapable of ever being bored! Nor can we take seriously Martin Gardner's attempt to explain how he might have relied throughout on trickery (Gardner, 1989). If he was, indeed, a trickster, he should have steadily improved as he became more practiced. Whatever the explanation of these long-term declines, it must surely be something deep and pervasive.

TWO FALSE DAWNS

One consequence of these historical vicissitudes is that, more than once, we seem to have stood on the threshold of a new era in which parapsychology would at last come of age and gain universal recognition. Two examples, in particular, impressed themselves upon me while I was writing my history. The first concerns the founding of the SPR in London in 1882; the second, the setting up of the Duke University Laboratory under Rhine. The beginnings of the SPR were modest enough. It was largely thanks to the financial support of the spiritualists, who initially made up much of its membership, that it could get going at all. But very soon it began attracting eminent scholars and scientists, plus many other notabilities of Victorian England (Gladstone, Tennyson, Ruskin, the painter G.F. Watts, and Lewis Carroll are among its earliest members). Within a decade it could boast no less than eight Fellows of the Royal Society on its Council. But far more important than its public relations was its sheer productivity in those pioneering days and the high standard of scholarship it sustained, as the early volumes of its *Proceedings* testify. This was mainly thanks to a few dedicated and indefatigable enthusiasts among whom Myers, Gurney, Hodgson, and Eleanor Sidgwick are perhaps the most outstanding. Their work includes the massive collection of spontaneous cases, the testing of mediums and psychics, laboratory experiments using hypnosis, and much else. Myers, in particular, enjoyed an international reputation and was prominent at the early international congresses of psychology. For, in the late nineteenth century, especially in France, there was considerable interest in the problems of dissociation, as exemplified by the various subconscious automatisms. This was a salient feature of Myers's own work with mediums, on the basis of which he built his theory of the subliminal self.

What went wrong? Why did this hopeful initiative fail to sustain its momentum? Why, indeed, has the membership of the Society never increased since those early days? We may get a clue when we turn to an article by William James which was first published in *The American Magazine* for October 1909 (Murphy & Ballou, 1960). James, I may say, was unquestionably the most distinguished individual ever to become President of the SPR (he was President from 1894–1895), although his greatest service to the field, on my reckoning, was to have discovered the incomparable Leonora Piper, whom he liked to call his "one white crow." She convinced him, beyond any scintilla of doubt, of the reality of psychical phenomena. At all events, in this 1909 article, he recalls an encounter with Henry Sidgwick, the first president of the Society, who died in 1900. I now quote his words:

> Like all founders, Sidgwick hoped for a certain promptitude of result; and I heard him say, the year before his death, that if anyone had told him at the outset that after twenty years he would be in the same identical state of doubt and balance that he started with, he would have deemed the prophecy incredible. It appeared impossible that that amount of handling evidence should bring so little finality of decision. (p. 310)

James then proceeds to discuss his own experiences, which, he tells us, have been similar to those of Sidgwick. "For twenty-five years," he tells us,

> I have been in touch with the literature of psychical research, and have had acquaintance with numerous "researchers." I have also spent a good many hours . . . in witnessing . . . phenomena. Yet I am theoretically no "further" than I was at the beginning; and I confess that at times I have been tempted to believe that the Creator has eternally intended this department of nature to remain *baffling*, to prompt our curiosities and hopes and suspicions all in equal measure, so that although ghosts and clairvoyances, and raps and messages from spirits, are always seeming to exist and can never be fully explained away, they also can never be susceptible of full corroboration. (p. 310)

James was not a pessimist or defeatist by nature, and I think his words reflect what many must have felt at that time although, compared with our own time, there seems to have been no dearth of exciting phenomena to study. The SPR was then embarking on its famous "Cross Correspondence" episode. This was only made possible by the availability of a whole galaxy of gifted automatists, most of whom were not professional mediums at all but educated, or even academic, women, such as Margaret Verrall and her daughter Helen, the like of whom we have never seen again. But, on second thoughts, James refuses to give way to despair. "It is hard to believe," he says,

> that the Creator has really put any big array of phenomena into the world merely to defy and mock our scientific tendencies; so my deeper belief is that we psychical researchers have been too precipitate with our hopes, and that we must expect to mark progress not by quarter-centuries, but by half-centuries or whole centuries. (p. 310)

Let us now jump several decades and consider the case of what I like to call the "Rhine Revolution." In 1933 J.B. Rhine produced a hastily written monograph which he called *Extra-Sensory Perception,* covering his researches of the previous few years. This he then sent to Walter Franklin Prince, head of the Boston SPR and one of his special mentors. Prince, however, could have had no idea what a hot property he was holding for, in 1934, he issued it in a limited edition of a mere 900 copies. The following year, however, Bruce Humphries of Boston, a commercial publisher, took it over, and the book was duly launched onto the general market. I think one can safely say that no other publication in the history of parapsychology or psychical research has ever achieved such acclaim or enjoyed such wide circulation. Science journalists in America wrote enthusiastic articles about Rhine for the national press, and, in departments of psychology up and down the United States, people applied themselves to this novel card-guessing test, which, as it happens, lent itself rather nicely to the fashionable behavioristic approach which American psychology was then everywhere adopting. If ever there was to be a breakthrough for parapsychology, this, surely, would be it. So what happened? What went wrong?

The brief answer is that the phenomena were simply not forthcoming. For Rhine, it was almost an article of faith that ESP was a universal endowment—we all had it to some small degree, and so everywhere it ought to show up as a statistical effect. Moreover, his initial experiences with ordinary student volunteers on the Duke Campus seemed to confirm his assumption. No doubt much of his early results can now be disregarded as due to the lax conditions of testing that then prevailed, but his monograph lists no less than eight named individuals who were capable of sustaining above-chance guessing under reasonable conditions of testing over a long enough period with astronomical odds against chance. These results cannot easily be set aside. Of the individuals involved the most outstanding was Hubert Pearce, a divinity student at Duke. Nothing, in my estimation, does more to confirm the genuineness of Pearce's performances than the ludicrous scenarios which Mark Hansel was driven to invent in order to discredit him.

Why, then, were there so few Hubert Pearces or even modestly proficient scorers on these other campuses? Why was Rhine himself soon unable to discover another Hubert Pearce? *I* do not know the answers to my questions, but I doubt if we can much improve on Rhine's own explanation as given in the introduction he wrote for the 1964 edition of *Extra-Sensory Perception.* There he tells us there was, at that time and place, a unique enthusiasm and a team spirit such as was never again to surface either there or elsewhere. If Rhine was right, he may, perhaps, furnish us with one of the clues to the repeated false dawns of parapsychological history to which I have drawn attention. For whatever reason—and we can return to that later—the manifestations of psi are, it seems, situation-dependent to a degree that has few parallels elsewhere in psychology. Indeed, I would go so far as to say that this conclusion may be the most important lesson of all.

The Rhine Revolution had two primary objectives. The first was to demonstrate to the satisfaction of the entire scientific community that the existence of ESP was a fact—for Rhine himself never wavered in his conviction that this was so. The second objective was to make parapsychology academically respectable. This he hoped he had achieved by taking the paranormal out of the murk of the seance room (his unfortunate brush with the medium Margery had left an indelible impression) and into the clear light of the laboratory. Rhine, I may say, worshipped science. For him it took the place of God once he had lost his religious faith as a young man (his mother had hoped he would become a minister). Yet, when Rhine died in February 1980, these two primary objectives were further than ever from fulfillment. On the one hand, the skeptical backlash had gathered momentum, and many eminent scientists were lending their support to a new skeptical organization, CSICOP, founded in 1976. On the other, popular occultisms and "New Age" fads were clouding the issue for the public at large and making it harder than ever for serious parapsychology to put its message across.

TWO SETBACKS

To add to his tribulations, two major scandals clouded Rhine's last years. The first, right on his doorstep, was the discovery that his protégé, Walter J. Levy, in whom he had reposed so much confidence that he had made him Research Director at his Institute for Parapsychology even before Levy had completed his medical studies, had been guilty of faking his data. This disaster, which could have destroyed the Institute, was all the more tragic inasmuch as Levy himself was, in my opinion, a highly gifted and ingenious experimenter besides being a tireless worker and enthusiast. Unhappily, owing to some weakness of character, he succumbed to the fatal temptation to cut corners and fabricate data when the desired results were not forthcoming. Ironically, his approach, which was based on automated testing of animals—he worked mainly with gerbils—might eventually have led to the kind of repeatable experiment that could have transformed the situation for parapsychology. For animals, after all, are not affected by those situational variables that make the testing of people so precarious. Although Levy forfeited all credibility, it would be unwise, I think, to ignore his own account of why he cheated. He was, he said, the victim of a decline effect; the success he had hoped for on the basis of his earlier work was not forthcoming, and he feared that he would lose the grant he was applying for.

The second scandal, which erupted a few years later (although it did not involve Rhine's entourage), was an even more serious disaster for parapsychology as a whole. I refer, of course, to the case of Samuel G. Soal, the erstwhile Mathematics Lecturer of London University and the leading exponent of experimental parapsychology in Britain. Rhine had little reason to be sorry for Soal, who had long been one of his harshest critics—nothing Rhine ever did could satisfy the perfectionist Soal. On the other hand, Rhine had gone on record as claiming that Soal's investigation of the card-guessing subject, Basil Shackleton, was one of the mainstays of the case for ESP—a view which I think most of us in parapsychology would then have endorsed, Curiously, there had been murmurings about Soal right from the time the testing of Shackleton was in progress in the early 1940s. One of Soal's collaborators, Gretl Albert, had declared that she had seen Soal altering figures and that Soal had had to silence her with the threat of legal action. It was not until much later, by which time Soal was too senile to speak for himself, that the late George Medhurst drew attention to suspicious-looking patterns in the data from some of the Shackleton sessions; his lead was followed up by Christopher Scott and Philip Haskell. But the final blow came only after Soal's demise when, in 1978, Betty Markwick published in the SPR *Proceedings* an account of her ingenious computer analysis of the data which revealed beyond all doubt that Soal had, indeed, been altering figures in the target sequence so as to produce spurious hits. Her discovery, I may say, was only possible because Soal had been in the habit of reusing his sequence of random numbers.

Perhaps one reason why Soal got away with it for so long is that it seemed incredible to many of us that he had eluded the watchful eye of his co-experimenter, the late and formidable Mollie Goldney. But more profoundly disturbing was how someone of Soal's caliber, who for so long had demonstrated his dedication to the field, could have stooped to such an act of treachery. Subject deception is, of course, something that we have all to learn to live with, however annoying and disruptive it can be. But for a scientist to falsify his or her findings is verily to sin against the Holy Ghost of Science. What possessed Soal to do it one can only speculate. It is known that he had a deviant personality—some would even go so far as to call it a split personality. Whether this goes back to his having been shell-shocked during the First World War, I do not know. I am inclined, however, to agree with Betty Markwick, who points out that Shackleton may well have been genuine when, at the instigation of Whately Carington, Soal re-examined his 1936 data for a displacement effect and thereby discovered his two celebrated subjects, Shackleton and Mrs. Stewart. "Then, as the ESP effect began to decline," says Markwick, "one may imagine that Soal resorted increasingly to boosting the scores." The thought of losing his first high-scorer after all those years of failing to find one may have overcome his normal scruples. If so, then we could say that here too, as with Levy, it was the ubiquitous decline effect that precipitated the tragedy.

THE PROBLEM OF FRAUD

Delinquent scientists are nothing new in history, and I do not think, in spite of the two highly publicized cases I have just mentioned, that parapsychology has attracted more than its share of such renegades. What is indisputable is that parapsychology is, for obvious reasons, so much more vulnerable than the conventional sciences to the adverse effects of such exposures. The only saving grace was that in both these instances, it was the parapsychologists themselves who exposed the offending practitioner. In the long run, however, what has done most damage to the image of parapsychology has been the confusion created by false claimants, those pseudopsychics who deliberately set out to deceive their investigators and sometimes get away with it. What should be our policy in the face of that danger?

Different conclusions have been drawn by different authorities regarding this particular painful lesson of history. Some parapsychologists, I know, are reluctant to take on board any purportedly gifted subject with a reputation to uphold, preferring to work with anonymous volunteers who make no special demands and insist on no special treatment but can be relied upon to submit meekly to whatever conditions are imposed—even if, at the end of the day, their scoring may be quite undistinguished. My own preference, however, is different. Believing as I do that a talent for psi is a very rare gift, but one well worth cultivating, I think we stand to lose more than we gain by ignoring it just because it is often so awkward to cope with.

No doubt, if one could say with assurance that a given individual is either genuine or a cheat, the situation would be much simpler. But history suggests that some of the most talented performers are what we have learned to call "mixed cases." We have already heard a good deal about Palladino at this Convention, and some of you may have read Richard Wiseman's recent article in the *Journal* of the SPR (Wiseman, 1992). But, in fact, nearly all physical mediums, from the Fox sisters onwards, were either detected in fraud or, like Eva Carrière or Margery, gave strong grounds for suspicion. One consequence of this is that the SPR, egged on by Hodgson and Podmore, decreed that physical mediumship was more bother than it was worth. Promising mediums, like Mrs. Leonard, were henceforth discouraged from becoming involved with it. As a result of this policy, most of the best work with physical mediums this century was done at the Institut Metapsychique in Paris. I have some sympathy with Brian Inglis, who accuses the leaders of the SPR of intellectual cowardice in this regard.

MATERIALIZATION

At the risk of shocking some of the more conservative among you, I would now like to say something on the vexed question of materialization. I think we would all agree that, if there really were such a thing as a full-form materialization, it would differ by another order of magnitude from any other class of paranormal phenomena known to psychical research. Nothing, after all, could be more incredible than the existence of a transient being who, for the duration of a seance, can talk, walk, and display, fully clad, all the functions of a corporeal human being. I would not blame anyone, therefore, for thinking that there is, in fact, no genuine instance of a full-form materialization. No one could deny, however, that there are, to say the least, some puzzling accounts in the literature that suggest otherwise.

A year after the famous Naples sittings, in October 1909, Hereward Carrington published in article in a popular American periodical, *McClure's Magazine,* entitled "Eusapia Palladino, The Despair of Science," and I am indebted to Richard Wiseman for bringing it to my attention. Palladino is not usually thought of primarily as a materializing medium. If she was able to make a small table levitate without using her feet or arms, or could make the curtains or her skirts billow in a closed room, her investigators were well pleased. Such materializations as she did sometimes produce were for the most part of a very degenerate kind, little more than stalks with knobs on. This, however, does not always appear to have been the case. Carrington relates the following curious anecdote about a man called Youriévitch who was a member of the Institut General Psychologique of Paris, which from

1905 to 1908 had carried out its own extensive, if somewhat inconclusive, series of tests with Eusapia. Some of the luminaries of the French intellectual scene, such as Pierre and Marie Curie and Henri Bergson, had been among the sitters. I now quote what Carrington has to say about this man, Youriévitch:

> His father had been dead for some years. At one of Eusapia's séances a solid though unseen body, tangible through the curtain, came to him, calling itself his father. Now his father had a peculiarly deformed finger: it tapered to a point, and the nail was deformed to suit the finger. M. Youriévitch asked his "father" in Russian—a language absolutely unknown to Eusapia—whether his father would impress his hand in the wet clay that was in the cabinet behind the curtain. Some time elapsed, the medium being carefully held and watched meanwhile. Soon the investigators were told to turn up the light and when they had done so and examined the clay in the cabinet, they found upon it the impression of a hand, the first finger of which bore identically the same marks of deformity as that of his long-dead father!

What is one to make of such a story? Is Carrington trying to put a fast one over on us? Would he risk being exposed to the world as a liar should someone pick him up on it? Or had Youriévitch perhaps invented the whole incident to see how much Carrington would swallow? But, in that case, what are we to say about Lombroso, the eminent Italian psychiatrist and criminologist? In a book dealing in part with Palladino, he reports that he witnessed the materialization of his own mother (Lombroso, 1909/1988, pp. 68–69).

He also informs us that Morselli, the director of a psychiatric clinic in Genoa who had carried out what was, I believe, the most comprehensive series of tests on Eusapia ever undertaken, which he then published in two volumes, had likewise encountered *his* mother at a seance, much to his own annoyance, since, according to Lombroso, Morselli was emphatically *not* a spiritualist. Lombroso also tells us that Bozzano once encountered his estranged wife at a seance. During her life, he had carried on a protracted litigation with her, and she was the last person he would have wanted to meet! Yet she spoke to him in a Genoese dialect which Eusapia, a Neapolitan, would not have known. Of course this is hearsay. We are not obliged to believe such stories, but in all honesty we must then ask ourselves why so many professional men should all have told the same sort of lie for no apparent reason!

I turn now to some much stronger evidence for full form materialization. In 1924 a certain F.W. Pawlowski, an American of Polish extraction who was then Professor of Aeronautical Engineering at the University of Michigan, went on sabbatical to Europe in the course of which he attended a seance in Warsaw given by the medium Franek Kluski. He describes his observations in an article he published in the *Journal* of the American SPR for September 1925. "Franek Kluski," we now know (see Weaver, 1992), was the pseudonym of Teofil Modrzejewski, a banker who was also a writer, a journalist, and a poet of some distinction. It was almost by accident, after attending a sitting with the medium Jan Guzik, that he discovered that, like his father, he had mediumistic gifts. He never used these gifts for mercenary ends even though he was, possibly, the most remarkable materializing medium of all time. One of the extraordinary features of his seances was the sheer number of assorted phantoms that simultaneously invaded the seance chamber, each performing different antics and often speaking different languages. Occasionally, the odd phantom bird or animal would also put in an appearance. Of course, all this took place in very subdued light, but many of the phantoms are described as being self-luminous. Still more bizarre, they often appeared first as less than life size, reaching normal adult size only gradually as the seance progressed and as the medium gained in strength.

Could Pawlowski have hallucinated the whole affair? It is, indeed, tempting to clutch at any straw. In fact, however, this is ruled out as a possible explanation because of the use of the following ingenious stratagem. It was customary at Kluski seances to ask the phantoms to dip their hands into liquid paraffin wax. As the wax cooled it would form itself into a glove—no more than about one millimeter thick. A human being cannot divest himself or herself of such a glove without shredding it, but a phantom, that can dematerialize, leaves an empty and unspoilt cast when it does so. This is what Kluski's phantoms obligingly did. Afterwards, plaster of paris can be poured into the empty glove and a plaster cast obtained showing the minute texture and skin markings of a human hand. Just to make sure that no substitute gloves could have been introduced at any stage, at some of these seances (notably those conducted by Geley in Paris) certain chemicals, such as cholestrin, were added to the paraffin wax without the knowledge of the medium or the sitters.

Fortunately some of these plaster casts have been preserved. Most of these are in Paris at the Institut, but I have seen some specimens at the SPR in London, and there are many illustrations in Geley's book (Geley, 1927/1975). Some of the Kluski casts are very complex, showing the two hands clasped together. Another curious feature of some specimens is that, though they have the proportions of an adult hand, they are the size of a child's hand. This nicely tallies with Pawlowski's observation that the phantoms only grew to normal size in the course of the seance. There is a passage towards the end of Pawlowski's article that is unbearably poignant in the light of all that has since transpired. For there he declares:

> I am perfectly convinced that we are on the threshold of a new science and probably of a new era. It is impossible for anyone to reject or to deny these phenomena, and it is impossible to explain them by clever trickery. I realize perfectly that it is difficult for anyone to accept them. (p. 503)

And a little later he adds:

> To accept them would mean to change entirely our attitude toward life and death, to be obliged to revise entirely our sciences and our philosophy. (p. 503)

We now know only too well what actually transpired. Kluski gave his last seance in 1926, after only seven years of mediumistic activity, and died in 1942 at the age of seventy. Nothing remotely like this was ever to be seen again. Today we pin our hopes, not on exceptional individuals like Kluski, but on the ingenuity and methodological excellence of our leading experimenters who, despite lack of funding, contrive to make steady, if unspectacular, progress. Yet I do not think I would be belittling current initiatives if I were to say that no one approach, however ingenious, is likely to go on paying dividends indefinitely. For if there is one lesson that our history should have taught us, it is that innovation is the name of the game and that routine is fatal to success.

CONCLUSION

If we adopt the skeptical interpretation and treat parapsychological history as nothing more than a succession of deceptions and blunders, then the decline of the phenomena can be easily explained. Every time serious investigators come to grips with some particular set of paranormal claims these become discredited, and the excitement dies down until a new set of impostors with a new repertoire of tricks comes to the fore and the cycle repeats itself.

The weakness of the skeptical interpretation is its failure to offer any specific, plausible, normal counter-explanation to the various episodes that go to make up our history. Until it does so, we must consider other possibilities, however far-fetched, for this "attenuation effect" in our history to which I have drawn your attention. I would like, therefore, to end by proposing one such idea.

Paranormal phenomena, we could say, represent a violation of the natural order. Now, nature reacts to such outrages much in the same sort of way that our bodies react to an infection. Hence, even if we can succeed in outwitting nature for a time, it will eventually get even with us, and we then have to start all over again on a new tack. If all this sounds too anthropomorphic for you, I must apologize; but allow me to develop my theme. Let us assume that the grosser this violation or infection, the more strenuously nature will strive to reassert the status quo.

Two implications can be derived from this model. First, we would predict that every new phenomenon or new approach in parapsychology is likely to flourish at first before nature has had a chance to rally her defenses. It will then gradually dissipate itself leaving in its wake doubt, confusion, and misgivings. For example, if crop circles are indeed paranormal (and I hazard no guess), we may surmise that, having reached a peak in number and complexity, they will then diminish in frequency and quality with each successive season until all that remains is the legend and a photographic record.

Secondly, however, we may suppose that the weaker the phenomenon in question, the smaller the departure from the norm which it represents, the longer it will endure. From our analogy with immunology we know that alien organisms, bacilli or whatever, can survive indefinitely in the tissues of a healthy body, provided only that they are sparse enough to cheat the immune system and pose no threat to the host. Hence we may predict a much longer life for the psi phenomena of the laboratory—although even there novelty should be at a premium.

Intuitively, one could try to order the variety of paranormal phenomena into a hierarchy according to the degree to which they upset the status quo. Thus, full-form materializations would mark the top end of the scale while micro-PK of the RNG variety would come at the bottom end. Good qualitative free-response ESP would occupy a position somewhere between these two extremes. The crucial point I want to make is that, no matter what the particular phenomenon we are dealing with, it is futile to hope that we could ever arrive at the perfect formula which, if adhered to faithfully, would guarantee a positive outcome. Repeatability in our field can never be absolute. Instead we shall have to recognize that we are caught in a battle of wits against the conservative forces of nature and that success will always depend on our being able to remain at least one step ahead.

These, at any rate, are the lessons I have drawn from my reading of our history. I can but pass them on to you for what they may be worth.

REFERENCES

Carrington, H. (1909). Eusapia Palladino: The despair of science. *McClure's Magazine*, *33*, 660–675.

Gardner, M. (1989). *How Not to Test a Psychic*. Buffalo, NY: Prometheus.

Geley, G. (1927/1975). *Clairvoyance and Materialisation*. New York: Arno. (Reprint of English edition, London: Unwin, 1927, translated from French edition of 1924).

Kappers, J., Akkerman, A.E., van der Sijde, P.C., & Bierman, D.J. (1990). Resuming Work with Pavel Stepanek. *Journal of the Society for Psychical Research, 56,* 138–147.

Lombroso, C. (1988). *After Death—What*? Wellingborough, Northamptonshire: Aquarian Press. (Original work published 1909).

Markwick, B. (1978). The Soal-Goldney experiments with Basil Shackleton: New evidence of data manipulation. *Proceedings of the Society for Psychical Research*, *56*, 250–277.

Markwick, B. (1985). The Establishment of Data Manipulation in the Soal-Shackleton Experiments. In P. Kurtz, (Ed.), *A Skeptic's Handbook of Parapsychology* (pp. 287–311). Buffalo, NY: Prometheus.

May, E.C., Humphrey, B.S., & Hubbard, G.S. (1980). *Electronic Systems Perturbation Techniques* (final report). Menlo Park, CA: SRI International.

Murphy, G., & Ballou, R.O. (1960). *William James on Psychical Research.* New York: Viking.

Pawlowski, F.W. (1925). The mediumship of Franek Kluski of Warsaw. *Journal of the American Society for Psychical Research, 19,* 481–504.

Radin, D.I., & Ferrari, D.C. (1991). Effects of Consciousness on the Fall of Dice: A Meta-analysis. *Journal of Scientific Exploration, 5,* 61–63.

Stevenson, I. (1990). Thoughts on the Decline of Major Paranormal Phenomena [SPR Presidential Address 1989]. *Proceedings of the Society for Psychical Research, 57,* 149–162.

Weaver, Z. (1992). The Enigma of Franek Kluski. *Journal of the Society for Psychical Research, 58,* 289–301.

Wiseman, R. (1992). The Feilding Report: A Reconsideration. *Journal of the Society for Psychical Research, 58,* 129–152.

Reaching for Consciousness: Expansions and Complements

William G. Braud (Institute for Transpersonal Psychology, 744 San Antonio Road, Palo Alto, CA 94303)

> Manifestation is mind; and so is Voidness too.
> Enlightenment is mind; and so is blindness too.
> —Tilopa, *The Vow of Mahamudra*
>
> Stop talking and thinking, and there is nothing you will not be able to know.
> —from the *Hsin-hsin-ming* ("On The Believing Mind") of Seng-ts'an, Third Patriarch of Zen

I initially thought of entitling this presentation "Reaches of Consciousness," then thought "Reaching for Consciousness" would be a more appropriate title—as a drowning person reaches for a life preserver, or as a growing plant reaches for the sun, or as a Seeker reaches for the Object of the Quest. And since Nature appears to abhor vacuums but adore complements, I felt it would be good to acknowledge and honor complements both in my title and in the content of my remarks.

COMPLEMENTS

In Nature and in our everyday lives, as well as in our science of parapsychology, balance is the key to proper functioning, well-being, and success. I believe a recognition and an appropriate honoring of complements is, in turn, a key to achieving balance. A complement is that which fills up, completes, makes perfect. Complements mutually supply each other's lack. Both are necessary to the completion and functioning of the whole. Each is incomplete without the other and, in fact, each only can be defined and only makes sense in terms of the other. Neither is more important than the other. To overemphasize the one, to ignore the other, is to miss at least half of reality, to encourage imbalance, to be unwise. An appreciation of complements is a healthful antidote to a limited and myopic view of the world. It replaces a quarrelsome, antagonistic, competitive, "either/or" view with a more agreeable, cooperative, "both/and" position.

The challenge of a complementary view is to learn to look beyond appearances, beyond polarities, to recognize that what seem to be opposites or contradictions or incompatibilities are not truly so. Each aspect is true and real and important under certain conditions. The challenge is to learn to live with and be comfortable with apparent contradictions. Physics has achieved this comfort with respect to the complementary wave/particle nature of light. Parapsychology can grow to appreciate and become comfortable with the complementary nature of its subject matter and of the various methods that have been proposed for its proper study.

Everywhere in Nature, complements abound—chaos and order, darkness and light, birth and death, assertiveness and surrender, activity and rest, emptiness and the Ten Thousand Things, mysticism and Las Vegas. Life's pendulum swings easily from one complement to the other, honoring each, never attaching too long to either extreme, never out of balance, ever leaving, ever returning. Life, as Nature itself, is a shifting rhythm of complements, a dance of opposites. So too, complementary approaches shall swirl and dance through this presentation.

I am speaking with you today as one who is interested in consciousness. Psi functioning is but one aspect of consciousness, and parapsychology is one of a number of complementary approaches to issues of consciousness. Parapsychology's findings regarding psi functioning remind us of those aspects of consciousness that do not appear to be limited by spatial, temporal, and material constraints as these are conventionally understood. They speak to us of the reaches of consciousness into remote, otherwise inaccessible recesses of space, time, and circumstance for purposes of direct knowledge, direct influence, and, no doubt, for other purposes that remain largely obscure.

In my homage to complements, let me begin by recognizing an obvious one in today's sociological realm—the advances and declines in parapsychology. On the one hand, we have stronger empirical evidence than ever (in the form of effective methodologies, extensive databases, impressive meta-analyses, and supportive review articles), rising public interest, and increasing media attention (in the form of articles, books, and television programs). On the other hand, we are experiencing reduced funding, greater academic and scientific resistances, decreased membership, reduced enthusiasm and excitement in our professional organizations, fewer contributions for our technical journals, and our laboratories closing down right and left. Even some of those whose doors remain open are facing troubling and sometimes painful reevaluations of how, and even whether, they plan to support parapsychological studies in future. To one familiar with Nature's ever-continuing dance of opposites or complements, such a situation is neither strange nor unexpected.

The chief message of my presentation is that we should expand our research endeavors. Expansions, provided they are wise rather than narrow or redundant, will necessarily allow us to confront increasing numbers of complements and broaden and deepen our appreciation of psi and of reality. I shall suggest several dimensions along which useful expansions may be possible. We can expand our experimental designs, the aspects of reality that we ask psi to address as we use these methods, the levels at which we ask psi to reveal itself, the accompaniments of our procedures, our subject matter, our methods of investigation, our sources of inspiration, our attitudes toward psi, our conceptualization of psi, our audience, and ourselves.

EXPANDING OUR EXPERIMENTAL DESIGNS

In recent years, experimental parapsychology has become synonymous, in large part, with Ganzfeld-ESP and random event generator (RNG)-PK research. I have viewed this development, alternately, as promising, expedient, and pragmatic and as limited, amusing, and troubling. And I say this as an early Ganzfeld researcher and one who has done his share of RNG-PK experiments. Let me suggest that we complement our Ganzfeld and RNG research with research on direct mental influence of biological systems (that is, what we used to call "biological psychokinesis" or "bio-PK" research). My motive for emphasizing this kind of research is selfish, for I have been engaged in this happy pursuit for many years.

In the English language scientific literature, there are approximately 100 published reports of experiments in which persons have been able to influence, mentally and at a distance, a variety of living target systems, including bacteria, yeast colonies, fungus colonies, motile algae, plants, protozoa, larvae, woodlice, ants, chicks, mice, rats, gerbils, cats, dogs, and dolphins, as well as cellular preparations (blood cells, neurons, cancer cells) and enzyme activity. In human "target persons," eye movements, gross motor movements, electrodermal activity, plethysmographic activity, respiration, and brain rhythms have been influenced. Much of this work has been reviewed by Jerry Solfvin (1984) and by Daniel Benor (1991).

The work is not really new. Rather, it represents conceptual replications and extensions of human distant mental influence experiments conducted many years ago (just before and just after the turn of the century) in France (by Joire, Gibert, Janet, and Richet) and in the former Soviet Union (by Bekhterev, Vasiliev, and their coworkers).

I do not have time to describe our direct mental influence procedures or results in any detail; for those interested, a very complete summary paper is now available (Braud & Schlitz, 1991). Suffice it to say that in our experiments on direct mental influence of remote biological systems, a variety of living targets have been used, including human electrodermal activity, muscular activity, and blood pressure; the spatial orientation of fish; the locomotor activity of small mammals; and the activity of *in vitro* cellular preparations (that is, rate of hemolysis—breakdown through osmotic stress—of human red blood cells). Using these eight target systems, we have conducted 37 experiments that have involved 655 sessions, 449 different "influencees," 153 different influencers, and 13 different experimenters. A statistical summary of these experiments yields a combined (Stouffer's) $z=7.72$ (which has an associated two-tailed probability of approximately 5×10^{-14}, and a mean effect size $(r)=.33$.

I recommend this direct mental influence of living systems paradigm for your consideration (and, hopefully, for your replication) for a number of reasons. First, the design has yielded positive and relatively consistent results. Conceptual replications of this work have been successfully carried out in independent laboratories in Brooklyn, NY, San Bernardino, CA, Freiburg, Germany, and Moscow, Commonwealth of Independent States (Kelly, Varvoglis, & Keane, 1979; Khokhlov, 1983; Gruber, 1979, 1980; Protosov, Baron, Druzhkin, & Chistyakova, 1983). Second, the experiments are easy to perform. Third, the living target systems are rich in free variability or *lability,* which has been shown to be favorable to the occurrence of psi effects. Fourth, the findings of such experiments are relevant to a potentially important psi application—mental healing (known also as absent healing, psychic healing, or spiritual healing). Fifth, the design is one that is rich in motivational components (which are accepted as critical to the success of psi experiments). Influencers who might not relate well to inanimate targets (such as random event generators) nonetheless enjoy the prospect of influencing other people or other living organisms, and the motivational levels of influencers, influencees, and experimenters alike are boosted by the recognition of the potential applications (especially for healing) and implications of the findings of these studies. Sixth, quantification of target activity is very straightforward and objective in these experiments (compared, for example, with the complexities of quantification by judges of results of free response ESP experiments). Seventh, multiple "psi channels" can be active and can contribute to a successful outcome in this design. There can be a direct "causal" psychokinetic influence by the influencer upon the influencee's biological activity. There can be an "active agent telepathy" effect by the influencer upon the influencee. Positive results can occur even if the influencer is not a psi source at all by the influencee learning the influence/noninfluence schedule via telepathy and combining this knowledge with physiological self-regulation in order to bring about the desired outcome; all of this can occur outside of the influencee's awareness. Similarly, the influencee can access the influence/noninfluence schedule clairvoyantly. To the extent that there is freedom in initiating a session or the trials of a session, an experimenter's precognitive waiting strategy or intuitive data sorting strategy could yield positive results; again, this could occur without the experimenter's awareness. Eighth, the task can be represented as one requiring joint effort by both influencer and influencee. The influencer reaches out to influence the influencee, but the influencee makes himself or herself open and receptive to such influence and reaches out (unconsciously and psychically) to determine when the influencer is making influence attempts and (unconsciously) letting the body respond appropriately. This shared responsibility for a successful outcome serves to reduce performance anxiety and decrease the psi-inhibitory effects of what the late Kenneth Batcheldor called "ownership resistance" (Batcheldor, 1984). And ninth, the experiment is a "general purpose" one that can be represented in different ways to different participants in order to match their interest and increase their involvement and motivation. For those interested in mind over matter effects, the task can be presented as one that involves a psychokinetic influence of the bodily systems of a distant person or other living organism. For those interested in healing, the task's healing analog nature can be emphasized. For those interested in telepathy, empathy, or connectedness, those aspects can be emphasized, with the physiological reactions serving as objective indicators or correlates of successful "merging" of influencer and influencee. I can think of no other psi task that is so rich in multipurpose elements.

Regrettably, time permits only passing reference to some of our findings. Physical factors of lability (free variability) and (perhaps) the momentary condition of the earth's geomagnetic field may relate importantly to the outcomes of direct mental influence sessions.

Physiologically, there are indications of greater susceptibility to electrodermal direct mental influence in influencees whose spontaneous electrodermal activity levels are relatively high, compared to those with relatively low activity levels. This is consistent with a hypothesis that living systems characterized by a greater degree of departure from homeostasis may be especially susceptible to direct mental influences aimed at restoring physiological balance. (I am thankful to Dr. Bruce Pomeranz for this observation.)

At the psychological level, we have learned that the ability to influence or to be influenced is relatively widespread among both selected and unselected participants. Participants enjoy the experiments, and few have indicated concerns about the possibility of influencing or being influenced by another person. Real time physiological feedback to the influencer is not essential to the occurrence of the effect. There are indications that the direct mental influence effect can be focused upon particular subsystems of the target through directed attention and intention, and that persons are able to "block" unwanted direct mental influences upon their own physiological activities. The physiological activities of a distant person can be influenced not only by directional intentional aims, but also by simply directing full attention toward the target person. The magnitude of a direct mental influence effect appears to be enhanced by the presence of an influencee's felt need to be influenced or by the influencer's perception of such a need in the influencee. Direct mental influences may occur in persons who are conventionally unaware that such influences upon their physiological activities are being attempted. I suspect, however, that even in such cases, there is a form of tacit knowledge and approval of the influence. Finally, we have found that the magnitudes of certain direct mental influence effects are correlated with influencee personality factors such as degree of introversion, degree of social avoidance and anxiety, and the degree to which one is comfortable and nondefensive about the possibility of being strongly "interconnected" with others.

Our direct mental influence work can easily be extended in several directions. There are already indications that animate target systems may be more susceptible to direct mental influence than are inanimate target systems (Braud, Schlitz, & Schmidt, 1989). It would be desirable, however, to conduct careful, analytical experiments in which animate and inanimate targets are compared directly, under identical psychological conditions. The outcomes of such comparison studies would have rich methodological, theoretical, and practical implications and applications.

In the experiments conducted thus far, the influencer's bodily and mental activities have been highly correlated. When I attempt to calm a distant person, I maintain an intention for calmness of the distant person, but I also calm and relax my own body. Which is the more critical factor? Is the physiological activity of the influencee "mirroring" that of the influencer and demonstrating a degree of physiological coherence? Or are the bodily reactions of the influencee more responsive to the influencer's *intentions?* What would happen if these two factors were aimed in opposite directions? If the influencer's intention were for calmness, yet the influencer's body were activated, which would the influencee follow? Would the intention override the physiological template, or would the opposite occur? Outcomes of such studies would have useful practical application.

How well can a direct mental influence be focused upon one of a number of bodily systems or upon one of a number of distant persons? And would the physical proximity of the persons play any role in such targeting attempts?

Can direct mental influence of living systems occur in a time-displaced manner? We are all aware of Helmut Schmidt's remarkable and successful experiments on retro-psychokinesis or time-displaced influences upon digital random event generators (Schmidt, 1987). Can similar, apparently "backward in time" influences be observed with analog, living target systems? Would persons be able to influence their own or others' prerecorded but not yet observed or inspected physiological activity? Would the degree of awareness of the response system (i.e., the degree to which the person is conscious or unconscious of the occurrence of the initial reactions) make a difference? This could be tested, for example, by comparing time-displaced influences upon relative unconscious reaction systems, such as electrodermal activity, with much more conscious reaction systems, such as muscular or motor activity. What would be the influence of pre-observations by others of the pre-recorded physiological activity? Would the time interval between the initial recording and the influence attempt be an important factor? One can imagine the possible medical implications of such experiments, especially in the areas of initial symptom formation and illness prevention. Can an individual's mental processes "reach backward in time" to actually influence the initial developmental probabilities of healthful or harmful physical changes? And would the degree of development of the changes, their degree of preobservation by self or others (by means of prior medical tests, for example), and their "age" have important bearings on the likelihood of success in influencing them? These may seem like implausible hypotheses, but they are no more implausible than psi itself and could readily be studied in careful experimental designs.

I mention these possible extension experiments because, without the necessary funding for this new research, I may not be able to follow up on such experiments myself; I am hoping, however, that some of you with the requisite facilities, equipment, and inclination may be able to conduct such studies and perhaps learn very interesting things.

EXPANDING OUR PSI INDICATORS

One of the interesting aspects of the research I have been describing is that the experimental design allows us to tap psi functioning at a relatively "unconscious" physiological level that does not require the generation of conscious, verbal reports. In a way, it was a kind of "historical accident" that led us to focus upon conscious, verbal reports as psi indicators. In the early years of psychical research, and in experimental parapsychology until quite recently, we could only know about a person's psi functioning if it took the form of conscious experience that could be discussed with the investigator. This is no longer the case. We now have physiological indicators of ever-increasing sophistication and subtlety that can be used to detect psi interactions that previously would have gone unnoticed. In our own work, for example, we have used unconscious motor automatisms successfully as psi indicators (Braud & Jackson, 1982). We have also observed positive results in sessions in which participants indicated their awareness of correct target events not by conscious guessing but by demonstrating greater unconscious electrodermal changes to correct targets than to nontarget decoys (Davis & Braud, 1980). Interestingly, in these sessions, accuracy was greater for the nonverbal, physiological indicators than for conscious, verbal guesses. The excellent results from these studies that employ nonverbal psi indicators as well as the relatively large overall effect size generated by our direct mental influence work (all of which employs nonverbal, physiological psi indicators) suggests that moving away from verbal to nonverbal indicators may indeed be a profitable research strategy. Gertrude Schmeidler's (1990) recent discussion of the essentially "subcognitive" or preconceptual nature of psi is relevant to this issue. Perhaps, when we attempt to tap psi at a physiological or preconceptual level, we may encounter fewer distortions or interferences by other processes, and our assessment may be somehow "purer" and less contaminated?

Others have already employed more sophisticated electroencephalographic, magnetoencephalographic, and averaged event-related potential measurements as useful psi indicators (e.g., May, Luke, Trask, & Frivold, 1990; Warren, McDonough, & Don, 1990). Even more sophisticated brain electrical activity mapping (BEAM) and positron emission tomography (PET) procedures may yet yield exciting results.

In addition to all of these physiological indicators, we now have a wide range of possible psi-mediated instrumental response (PMIR) indicators of psi, thanks to the thinking and work of Rex Stanford (Stanford, 1990). These PMIR indicators can now allow us to detect instances in which organisms may unconsciously scan their environments for need-fulfilling interactions and change the timing or pattern of reactions in order to increase the likelihood of such interactions. A bonus of the PMIR methodology is that it allows us to conduct otherwise impossible psi investigations with animals, preverbal infants, or nonverbal adult participants.

It should also be possible for us to extend the range of psi indicators in the other direction, so to speak—to tap still other levels of psi responsivity. The direct mental influence methodology, for example, could be extended from biological "target systems" to psychological, social, and even to psi-functioning targets. We are engaged in experiments in which we are exploring whether one person's attention-focusing processes can influence another person's simultaneous attention-focusing or concentration attempts; stated somewhat differently, can one person's meditation practice directly facilitate that of another, distant person? Here, of course, the target system is a psychological one.

We are also currently exploring still another psychological target system in an experiment in which persons attempt to influence the physical and psychological health and well-being of another person through the vehicle of intercessory prayer. This multiple-blind prospective study is a conceptual replication and extension of a similar experiment reported in the *Southern Medical Journal* by cardiologist Randolph Byrd (1988), in which he claimed to find dramatic intercessory prayer effects in cardiac care patients. For us, this study represents three extensions of our ongoing direct mental influence work: (a) from laboratory to clinic (the influencees are clients experiencing counselling at a local pastoral counseling center), (b) from physiological to psychological targets, and (c) from an egocentric, influential mental strategy to a less egocentric, prayerful mental strategy. Persons contemplating such studies should expect the need to deal with many levels of resistances, which we experienced but which we finally have been able to overcome; the study is now proceeding smoothly.

Social processes could provide targets for direct mental influence attempts. These target systems could range from the social activities of community organisms such as ants or bees that exhibit hive behaviors to simple and complex social behavioral and decisional processes in dolphins and in humans. Societal target experiments have been reported by investigators of Transcendental Meditation (e.g., Orme-Johnson et al., 1988), and while such experiments remain highly controversial and their interpretation unclear (e.g., Schrodt, 1990), the political, economic, and societal domains in which the experiments and controversies are being conducted and played out indicate the potential importance and practical impact of such work.

EXPANDING OUR "TARGET EVENTS"

The foregoing suggestions were extensions of the direct mental influence design. Of course, other experimental methods can be extended as well and may yield new and exciting results. Consider the Ganzfeld. Our present use of the Ganzfeld is paradoxical and ironic in that we use it as a means of reducing external distractions and directing attention inwardly; that is, we use the technique to release the participant from the world of sense objects. But we direct this newly freed attention to images of pictures, to the world of sense objects one step removed. This may not be the best use of the technique. We free the mind of its senses (chiefly vision) so that the mind can use its psi abilities as new senses (chiefly vision). What would happen if we were to allow the sense-freed mind to do other things? As an easy first step, we could employ sense object-like targets that are nonvisual. Auditory, gustatory, olfactory, and tactile target events could be used. As Lawrence LeShan (1977) has suggested, perhaps individual differences will blossom here, with participants showing different sensitivities or preferences for targets of different sense modalities. Next, we could expand to more "internal" targets—for example, proprioceptive or visceral stimulation of another, distant person. And beyond that, we could leave sense-type targets behind in order to explore target events that consist of different emotions, different cognitive processes, or different states of consciousness of other persons. More interestingly still, we could investigate what I termed "nonevident psi" in an earlier paper (Braud, 1982). Here, we could begin by assuming that Nature would not have produced a psi process that was completely redundant with sensory or perceptual processes and ask what psi might be able to tell us about the world that cannot readily be discerned by the conventional senses. Perhaps psi is more adapted to providing knowledge about objects that would not be available to the senses themselves—such as the origin, history, purpose, associations, relationships, interconnections, or meaning of objects. William Roll (1982) has made related suggestions.

Even more interestingly, we might encounter an entirely new realm that is not at all sense-like and that is difficult or impossible to describe in ordinary language. What would happen if we prolonged and restructured our Ganzfeld procedures and instructions so that the mind could go not only beyond sense impressions, but beyond images as well—beyond the noise at one or several removes to silence to the absence of sensory, emotional, cognitive, and intentional content; beyond anticipations; beyond memories; beyond the visual vehicles of psi? What might we discover? How might we meet the challenge of developing and utilizing new means of experiencing, describing, representing, understanding, and communicating this reality? Might we discover that some of these are inappropriate and that the realm must necessarily remain cloaked in a cloud of unknowing? Or will we break through to new levels of knowing and even being, for which we will find increasing intersubjective verification that our sophisticated experimental designs can register?

Approaches to this realm are, of course, not new. Charles Tart called our attention to the general problem area in his seminal "state-specific sciences" paper that appeared in *Science* two decades ago (Tart, 1972), and partakers of the mystical experience have been attempting to describe this conceptless, wordless realm with words and concepts for millennia (Underhill, 1911). To my thinking, the mystics have been amazingly consistent and successful in this undertaking, and the Way of the Mystic may in fact be one of the state-specific sciences envisioned by Tart. The Way of Meditative Discipline may be another such state-specific "mind science." Still other examples may be found if one looks cautiously at the various esoteric and mystery systems, carefully discarding the associated nonsense that one finds with these, although such distinctions are not always easy to make!

A familiar image, with which all of us are doubtlessly familiar, is the San-en or Three Monkeys: See No Evil, Hear No Evil, and Speak No Evil. We have already followed the advice of the first two monkeys (Mi-saru and Kika-saru: Don't see and Don't hear) in the visual and auditory restriction portions of the Ganzfeld procedure. We have yet to adequately heed the advice of the third monkey (Iwa-saru: Don't speak). Let me suggest that we introduce more of the nonspeaking, the nonverbal, into our experimental designs and into our other means of investigation. Silence may be a useful complement to our more familiar ways of acquiring knowledge. Time spent in wordless, conceptless silence, in our experiments and in our nonlaboratory experiences, may yield unexpectedly rich rewards of enhanced psi, enhanced creative insights, and other gifts of knowledge and inspiration.

OTHER ACCOMPANIMENTS OF OUR PROCEDURES

Let us continue with the Ganzfeld. That procedure has been used within experimental parapsychology as a facilitator of free-response GESP. There have been a few exceptions—for instance, Rex Stanford's investigations of the cognitive concomitants of Ganzfeld exposure (Stanford, 1987); a master's thesis (Stembridge, 1972) which I directed on Ganzfeld-correlated enhancements of creativity and originality; our studies, reported in 1985, of physiological concomitants of Ganzfeld exposure (Braud, Davis, & Opella, 1985); and some very recent creativity-related investigations by Marilyn Schlitz and Charles Honorton (Schlitz & Honorton, 1992)—but the Ganzfeld has been used almost exclusively as a facilitator of accurate telepathic impressions of visual target materials. It is somewhat embarrassing to add that we are not even certain that the Ganzfeld itself is a *facilitator* of such visual target telepathy, for (with only a handful of exceptions) simple, non-Ganzfeld control groups are not even used in the experiments. So what are the Ganzfeld meta-analyses really telling us? Are the obtained positive results due to the Ganzfeld? Or are we simply being told that psi often comes out to play in certain laboratories or with certain investigators, regardless of the techniques used?

It would be easy to add control groups to our Ganzfeld experiments. And it would be easy to study the Ganzfeld as a possible facilitator or amplifier of a far greater range of processes. Jeanne Achterberg has reminded us that throughout history, most, if not all, natural healing systems (from ancient dream incubation procedures at Asclepian temples to modern techniques of hypnotherapy and psychoimmunology) make use of images presented to patients who are in a state near sleep (Achterberg, 1985). The Ganzfeld has exactly these two characteristics—imagery becomes very prominent, intense, and realistic, and the participant enters a hypnagogic or twilight state of consciousness (Mavromatis, 1989; Schacter, 1976). This happy circumstance suggests that the Ganzfeld might provide a very simple and inexpensive method for facilitating health-enhancing imagery and other health-enhancing processes. We are, in fact, currently pursuing the possibility of using the Ganzfeld as a facilitator or amplifier for processes helpful to psychophysiological self-regulation, self-healing, and direct mental influence of animate and inanimate target systems. The Ganzfeld might also be used to increase the susceptibility of influencees to the types of biological psychokinesis effects I discussed earlier.

Parallel investigations could be carried out using other ostensibly psi-favorable techniques, such as relaxation, meditation, special forms of intentionality, or simply dwelling more frequently in or attending more to the silence between thoughts. In these, and in Ganzfeld investigations, physiological and experiential concomitants could be studied during sessions in which participants attempt to become aware of remote target events. Similar studies also could be undertaken during sessions in which there were no specified targets, in order to learn just what psi might "capture" when not directed to a particular type or target "location," as well as to learn whether the physiological and experiential concomitants differ for sessions with and without concurrent psi attempts. The "targetless" experiments could emphasize a more general understanding of the techniques and of the broad range and nature of experiences facilitated by the techniques, quite apart from their real or ostensible "paranormal" components.

EXPANDING OUR SUBJECT MATTER

Along similar lines, we could expand our subject matter beyond the usual "big four" processes of telepathy, clairvoyance, precognition, and psychokinesis. I am not alone in suggesting such an expansion of the issues, topics, and subject matter of our field. In "The Limits of Parapsychology," John Beloff (1978) cautioned parapsychologists not to rule out prematurely certain unusual phenomena that might fall outside the traditional psi domain—that we at least be willing to explore anomalous manifestations other than ESP and PK. In his 1979 presidential address to the Parapsychological Association, John Palmer suggested that, in view of the progressive decline in "transmission" models and the growth of "correspondence" models of psi, parapsychology might expand its focus to consider additional "correspondences" of the types found in astrology, synchronicity studies, and in the magical tradition (Palmer, 1980). Susan Blackmore (1985, 1988) has often argued for the expansion of parapsychology to include unusual mental phenomena, such as ostensible psi experiences but also such things as out-of-body experiences, near-death experiences, lucid dreams, mystical experiences, and altered states of consciousness that seem to suggest other "planes" of mental functioning, and that parapsychology study these mental states in their own right, whether or not they happen to be accompanied by veridical psi. I agree with Blackmore's suggested expansion of parapsychology in this way; I disagree, of course, with her rejection of veridical psi phenomena and with her suggestion that we abandon the psi hypothesis as a blind alley. And most recently, Rhea White (1990, 1991) has proposed that we return to the study of spontaneous experiences and broaden our field by including studies not only of psi experiences but also other exceptional human experiences such as mystical and peak experiences, experiences of inspiration, out-of-body experiences, near-death experiences, UFO encounters, and accompaniments of exceptional human performance, because psi is often reported to be associated with them or increased after them.

It may be useful to reiterate my opening comments on complements at this point. I am not suggesting that we substitute a psi-less parapsychology for one that contains psi, nor am I suggesting that we abandon an experimental approach in favor of an experiential approach. We can do both—*complement* our work with related, nonpsi studies and *complement* our experimental work through the addition of more experiential and spontaneous investigation aspects. Not "either/or," but rather "both/and." These are complementary approaches, or perhaps the poles of continua; each can illuminate and enhance our study of the other, and the approaches can join and transform one another in a cooperative and synergistic, rather than a competitive and exclusive, manner.

EXPANDING OUR METHODS OF INVESTIGATION

Would it be helpful to consider expanding our methods of investigation as well? Assuredly, we can move in the direction of improved technology—tighter experimental designs, improved statistical techniques, more impressive apparatus, experiments that examine geomagnetic field interactions, protocols involving high tech brain mapping, new and improved random event generators, and increasingly analytical investigations of possible physical, physiological, and anatomical correlates or influencers of our studied phenomena. No doubt such a strategy is appealing to many, can lead to important insights, and may even attract funding. But to others, to whom the specific "hardware" through which events come about is less important than the events or processes themselves and their psychological natures, such technological adventures soon lose their interest.

One can move in other, complementary, directions as well: toward applying what we know or suspect and noting the results of those applications; toward an increased appreciation of the meaning of various experiences; toward a more complete understanding (see Abel, 1953) as well as explanation of experiences; toward exploring diverse sources for findings and interpretations that might illuminate what we are studying; and toward increased self-work, self-observation, and self-understanding. I shall have more to say about these later.

EXPANDING OUR SOURCES OF INSPIRATION

In addition to more extensive surveys of our own literature of psychical research and parapsychology, increased study of the literature of other scientific disciplines can uncover methods, findings, and concepts of great usefulness in our studies of psi. For example, in Part 2 of her journal, *Exceptional Human Experience,* Rhea White covers over 300 abstracts, many of them taken from other disciplines. But we can also range more freely and widely in our search for data, information, approaches, and interpretations that may help us in more fully understanding and appreciating our subject matter. We may find useful clues in unexpected places such as the mystical tradition, various meditative traditions, esoteric teachings, mythology, and even art, poetry, and music. To the extent that much of what we wish to study is nonverbal, it is not surprising that nonverbal modes of expression, such as music or art, could be quite revealing and informative. Concerning the essentially nonlinguistic nature of psi, Ivor Grattan-Guinness has provided many useful clues (Grattan-Guinness, 1985).

A specific illustration will provide an indication of what I am proposing. Consider this quotation:

> There is within us an immense capacity for perception, for the receiving of messages from outside; and a very little consciousness which deals with them. It is as if one telegraph operator were placed in charge of a multitude of lines: all may be in action, but he can only attend to one at a time. In popular language, there is not enough consciousness to go round. Even upon the sensual plane, no one can be aware of more than a few things at once. These fill the centre of our field of consciousness: as the object on which we happen to have focussed our vision dominates our field of sight. The other matters within that field retreat to the margin. We know, dimly, that they are there; but we pay them no attention and should hardly miss them if they ceased to exist.
>
> Transcendental matters are, for most of us, always beyond the margin; because most of us have given up our whole consciousness to the occupation of the senses, and permitted them to construct there a universe in which we are contented to remain. Only in certain states—recollection, contemplation, ecstasy and their allied conditions—does the self contrive to turn out the usual tenants, shut the "gateways of the flesh," and let those submerged powers which are capable of picking up messages from another plane of being have their turn. Then it is the sense-world which retreats beyond the margin, and another landscape that rushes in . . . by a delib-

> erate inattention to the messages of the senses . . . [one] can bring the ground of the soul, the seat of "Transcendental Feeling," within the area of consciousness: making it amenable to the activity of the will. Thus becoming unaware of his usual and largely fictitious "external world," another and more substantial set of perceptions, which never have their chance under normal conditions, rise to the surface. Sometimes these unite with the normal reasoning faculties. More often, they supersede them. Some such exchange, such "losing to find," appears to be necessary, if [one's] transcendental powers are to have their full chance.

The writer then goes on to describe methods of quietude, focusing of attention in a one-pointed manner, focusing of "naked intention," and so forth, as ways of attaining a special, different way of knowing. To me, these statements do not sound very different from the familiar statements of the "noise-reduction" models of psi optimization that Charles Honorton and I have proposed (Braud, 1978; Honorton, 1977). The statements were written in 1911 by Evelyn Underhill in her classic study, *Mysticism: A Study in the Nature and Development of Man's Spiritual Consciousness*.

Two interpretations may be given regarding the similarity of recent and older and even ancient thoughts. A pessimistic one is that we have not progressed very far from those olden times and thus know relatively little even today. A more optimistic assessment is that even the ancients possessed a great deal of valid and useful information, which they learned quite early and informally from firsthand experiences, ponderings, and appreciations, and that we have much to learn from these ancients even today. Such knowledge may be similar to that of the very young child who has acquired incredibly accurate knowledge of the physics of trajectories through experience in tossing and catching balls and stones; this is an informal, tacit knowledge that has not yet been communicated to others nor expressed in mathematical formulae. If the ancients were correct in some of their thoughts and writings, perhaps we could find many more useful findings, concepts, and interpretations were we to delve more widely and deeply into their writings.

Many of us have dipped deeply into such ancient sources. I do not know how widely appreciated it is that Charles Honorton's noise-reduction model (Honorton, 1977) and, in fact, his original

Ganzfeld experiment were inspired to a large extent by his readings of Patanjali's yoga sutras (dating from the second century B.C.). Some of Rex Stanford's ideas about the interference of cognitive and other constraints owe a debt to similar sources (Stanford, 1990), and K. Ramakrishna Rao has advocated that we research the classical texts of the meditative tradition and carefully study the phenomenology of adept meditators (Rao, 1989). It is also no secret that Lawrence LeShan's methods for training healers owe much to the Western mystical tradition (LeShan, 1974), that William Roll's "long-body" has much in common with the notion of the Mystical Body of Christ (Roll, 1988), and that Dolores Krieger's "therapeutic touch" is a mixture of the ancient practice of laying on of hands and of old Mesmeric procedures (Krieger, 1979). Nor, to give but one example from outside of our field, is it usually appreciated that the powerful notions of general system theory proposed and developed by Ludwig von Bertalanffy (1968) were largely inspired by von Bertalanffy's interest in the ideas of the mystic Nicholas of Cusa (best known for his doctrine of the coincidence of opposities in the ultimate reality and about whom von Bertalanffy wrote his own doctoral dissertation in 1928). The lesson is that there is much latent wisdom in ancient traditions. Combining this ancient wisdom with modern methods of gaining knowledge will yield a fuller, more complete understanding of consciousness, human nature, and reality than that gained by either one or the other alone.

EXPANDING OURSELVES

Another helpful expansion of our studies would be to engage in more self-work, self-development, and self-observation. I can give you a flavor of what I am suggesting by sharing several quotations that I chanced upon while preparing this address. "We behold that which we are," "Only the Real can know Reality" (both of these from Evelyn Underhill, 1911), and "No man can create anything greater than his own soul" (from a discussion of Robert Browning's poem "Andrea Del Sarto" in Witherspoon, 1951, p. 1013). Similarly, there are many relevant quotations in the writings of Gerald Heard (which I am indebted to Rhea White for bringing to my attention), all to the effect that "being determines cognition"—for example, "To listen *to* the soul, they must listen *with* the soul"; "If we change ourselves, changing our power of apprehension, we change the universe confronting us"; and "Every insight into the outer world must be balanced by an equally enlarged knowledge of [our] true and full nature" (White, 1982). What I mean to imply by these quotations is that a most effective way to learn about consciousness and about psi is to observe, develop, and study our own consciousness, in a first-person participatory manner, as a complement to the usual third-person observer manner characteristic of Victorian science. Again, I am merely echoing advice already offered by several psi investigators. Perhaps the first formal advocacy of this idea was by philosopher and parapsychologist C.J. Ducasse, who suggested, "Some persons disciplined in science and philosophy should strive painstakingly and persistently to develop paranormal abilities in themselves" (Ducasse, 1962). This has been proposed more recently: for example, by Rex Stanford in his 1973 presidential address to the Parapsychological Association (Stanford, 1974), wherein he suggested: "Experimenters should endeavor personally to experience as many psi phenomena as they can. They may also profit by being subjects in experiments of their own or similar to their own." In the same address, Stanford advocated the study of the practices and beliefs of magic, religion, and mysticism in diverse cultures and times as they relate to possible psi phenomena. Similar participatory and experiential suggestions have been made by Hoyt Edge (1982), David Hess (1988), Rhea White (1987, 1990, 1991), and others. What we learn and become in the process of self-work can be applied to and can facilitate our scientific work. Similarly, our laboratory and field findings can provide helpful hints for more effective and more efficient self-work. Again, a synergistic, complementary relationship—"both/and" rather than "either/or."

EXPANDING OUR CONCEPTUALIZATION OF PSI

We have several alternative conceptualizations of the nature and purpose of psi, ranging from an erratic flaw in our filtering equipment, to a primitive communication system, to a biologically adaptive ability, to a mental faculty, to an indication of our interconnectedness with one another and with all things, to the "glue" that holds together, balances, and orders Nature, to a means of interacting with a transcendent realm. Undoubtedly, reality is even more complicated than this, and psi may have additional functions and purposes about which we are still ignorant. Perhaps psi is an indication that our conceptualizations themselves are limited and incomplete and that there are realities that can never be reduced to conceptualizations. Our complementarity principle reminds us that we need not choose any one conceptualization of psi to the exclusion of others. Psi can be several of these things, and more, all at once.

A related area for expansion is our view of the range of what is possible or impossible to accomplish via psi. If psi accomplishments are indeed influenced by our expectations and beliefs, perhaps by expanding our belief systems, we could increase the range and magnitude of our psi effects, especially our PK effects. Many of the commonly assumed limits of psi may be just that—self-imposed and limiting assumptions. By fully suspending disbelief, fully encouraging positive beliefs and expectations, we may be able to accomplish formerly unthinkable feats via ESP and PK. For example, are PK effects *really* limited to small scale shifts of probabilities of random events? If we assume this to be the case, we shall never venture forth far enough or frequently enough to attempt more outrageous, yet perhaps truly possible, feats including large scale macro-PK effects. We shall never know what is possible or impossible until we give such attempts fair trials. And I mean truly fair opportunities, not half-hearted attempts or attempts that are not sufficiently thorough or intensive.

Similar comments apply to our assumptions about the roles of various physical, physiological, and psychological factors or variables in our experiments. Many of our supposedly discovered laws may in fact be self-generated and self-fulfilling, self-maintaining prophesies. We shall only know if this is so if we are able to go beyond our usual assumptions and test reality on its own terms—fully and sufficiently frequently. For example, nearly all of us take for granted the importance of immediate, trial-by-trial feedback in the production of psi interactions and go to great lengths to build this into our experiments and our models and theories. But there are many cases, in lab and in life, in which excellent psi appears to occur without such feedback. I have heard of a case in which a particularly accomplished remote viewer provided an extremely accurate and detailed remote viewing protocol—then died before receiving feedback. This would seem to be an excellent white crow that disproves the "all black crows" rule that feedback is necessary for effective psi functioning. This is merely one example among many that could be given regarding many "crucial" variables that may not really be crucial. Expanding beyond our assumptions regarding any given variable or process may yield wonderful surprises.

EXPANDING OUR AUDIENCE

Another area of possible expansion is the audiences that we address. In our work and in our pronouncements as parapsychologists, to whom are we speaking? I think we speak, in gatherings such as this, chiefly to ourselves and to our peers. Secondly, we speak to our critics. Thirdly, we speak to scientists at large. But there is a vast group that we usually ignore, and that is the public at large. What about the public? What can we offer them? And what can they offer us? At a recent Symposium on Science and Consciousness held in Athens, Greece, which I was honored and delighted to attend, one of the organizers made the following statement in addressing the delegates (scientists and consciousness researchers) during the very last moments of the conference: "You are the modern shamans, the modern priesthood, but if you do not bring your knowledge to the people, there will be barbarism." It is important for us, and for our world, to share the knowledge and wisdom that we have with the public. They are our greatest supporters; they are in need of any insights that can improve their lives; and they possess a great storehouse of knowledge that they can share with us, in return. Certainly, we can continue to speak among ourselves, and to our peers, our critics, and other scientists; but we can make greater efforts to address the general public and to listen to and honor what they have to say to us, in turn.

EXPANDING OUR ATTITUDE TOWARD PSI

Another area for expansion is in the way in which we approach our subject matter, our attitude toward psi. Along with our aim of explaining, predicting, and controlling psi, would it be helpful to add the complementary approach of honoring and appreciating psi without wishing to capture it or force it or constrain it? A member of our organization, a "tough-minded experimentalist" whom we all very much respect, shared with me at an informal gathering several years ago that he tended to think of psi as a beautiful painting to be admired, not something to be distorted or destroyed through overanalysis, or in the service of technological explanation. A similar point was made in an article by Joseph Friedman (1984) called "Love and Parapsychology," in which he argued that it may not be appropriate to treat psi as something to manipulate, will, and harness for our own designs; rather, psi may be part of that order of human experience that includes wisdom, humility, and love—an order "not subject to an approach which challengingly sets upon nature to order its being through will or knowledge."

Perhaps I was describing, to myself, two complementary approaches to psi research when I wrote these two *haiku* (which I present to you with great apologies to Basho, Buson, Issa, Shiki, and all other true practitioners of this wonderful Japanese poetic form). The first was written fifteen years ago, and it, like the method it describes, is strained and ineffective:

Catching a greased pig
with gloves filled with quicksilver:
  psi experiments.

The second, much more effortless, virtually wrote itself about eight months ago:

Seeds fall to the soil...
Rains fall, day replaces night...
  Butterflies arrive!

And perhaps this may serve as a useful recipe for a successful psi experiment, regardless of its specific methodology. We can set the stage, remove the things that get in the way, wait patiently for psi's natural coming, and experience delight when it arrives.

REFERENCES

Abel, T. (1953). The Operation Called *Verstehen.* In H. Feigl & M. Brodbeck (Eds.), *Readings in the Philosophy of Science* (pp. 677–687). New York: Appleton-Century-Crofts.

Achterberg, J. (1985). *Imagery In Healing: Shamanism and Modern Medicine.* Boston: New Science Library.

Batcheldor, K. (1984). Contributions to the Theory of PK Induction from Sitter-group Work. *Journal of the American Society for Psychical Research, 78,* 105–122.

Beloff, J. (1978). The Limits of Parapsychology. *European Journal of Parapsychology, 2,* 291–303.

Benor, D. J. (1991). Survey of Spirtual Healing Research. *Complementary Medical Research, 4,* 9–32.

Bertalanffy, L. von (1928). *Nikolaus von Kues.* Munich, Germany: G. Muller.

——— (1968). *General System Theory.* New York: Braziller.

Blackmore, S. (1985). Unrepeatability: Parapsychology's Only Finding. In B. Shapin & L. Coly (Eds.), *The Repeatability Problem in Parapsychology* (pp. 183–206). New York: Parapsychology Foundation.

——— (1988). Do We Need a New Psychical Research? *Journal of the Society for Psychical Research, 55,* 49–59.

Braud, W. G. (1978). Psi Conducive Conditions: Explorations and Interpretations. In B. Shapin & L. Coly (Eds.), *Psi and States of Awareness* (pp. 1–41). New York: Parapsychology Foundation.

——— (1982). Nonevident psi. *Parapsychology Review, 13*(6), 16–18.

———, Davis, G., & Opella, J. (1985). Electrodermal and imagery concomitants of Ganzfeld stimulation. *Parapsychology Review, 16*(4), 1–5.

———, & Jackson, J. (1982). The Use of Ideomotor Reactions as Psi Indicators. *Parpsychology Review, 13*(2), 10–11.

———, Schlitz, M. J., & Schmidt, H. (1989). Remote Mental Influence of Animate and Inanimate Target Systems: A Method of Comparison and Preliminary Findings. *Proceedings of Presented Papers: The Parapsychological Association 32nd Annual Convention,* 12–25.

———, & Schlitz, M. J. (1991). Consciousness Interactions with

Remote Biological Systems: Anomalous Intentionality Effects. *Subtle Energies, 2,* 1–46.

Byrd, R. (1988). Positive Therapeutic Effects of Intercessory Prayer in a Coronary Care Unit Population. *Southern Medical Journal, 81,* 826–829.

Davis, G., & Braud, W. [G.] (1980). Autonomic "Recognition" of ESP Targets. In W. G. Roll (Ed.), *Research in Parapsychology 1979* (pp. 58–60). Metuchen, NJ: Scarecrow Press.

Ducasse, C. J. (1962). In: The future of Parapsychology: A Symposium. *International Journal of Parapsychology, 4*(2), 5–26.

Edge, H. (1982). Some Suggestions for Methodology Derived from an Activity Metaphysics. In B. Shapin & L. Coly (Eds.), *Parapsychology and the Experimental Method* (pp. 43–64). New York: Parapsychology Foundation.

Friedman, J. (1984). Love and Parapsychology. *Journal of the Society for Psychical Research, 52,* 253–260.

Grattan-Guinness, I. (1985). Is Psi Intrinsically Non-Linguistic? Some Preliminary Considerations. *Journal of the Society for Psychical Research, 53,* 8–17.

Gruber, E. (1979). Conformance Behavior Involving Animal and Human Subjects. *European Journal of Parapsychology 3,* 36–50.

——— (1980). PK Effects on Pre-recorded Group Behavior of Living Systems. *European Jornal of Parapsychology 3,* 167–175.

Hess, D. (1988). Gender, Hierarchy, and the Psychic: An Interpretation of the Culture of Parapsychology. *Proceedings of Presented Papers: The Parapsychological Association 31st Annual Convention,* 341–353.

Honorton, C. (1977). Psi and Internal Attention States. In B. B. Wolman (Ed.), *Handbook of Parapsychology* (pp. 435–472). New York: Van Nostrand Reinhold.

Kelly, M., Varvoglis, M., & Keane, P. (1979). Physiological Response During Psi and Sensory Presentation of an Arousing Stimulus. In W. G. Roll (Ed.), *Research in Parapsychology 1978* (pp. 40–41). Metuchen, NJ: Scarecrow Press.

Khokhlov, N. (1983). Remote Biofeedback in Voluntary Control of Heart Rate. *Psi Research, 2,* 66–92.

Krieger, D. (1979). *The Therapeutic Touch.* Englewood Cliffs, NJ: Prentice Hall.

LeShan, L. (1974). *The Medium, the Mystic, and the Physicist.* New York: Viking.

———(1977). The Purpose of Psi. *Journal of the Society for Psychical Research, 49,* 637–643.

Mavromatis, A. (1987). *Hypnogogia.* New York: Routledge & Kegan Paul.

May, E., Luke, W., Trask, V., & Frivold, T. (1990). Observation of Neuromagnetic Fields in Response to Remote Stimuli. *Proceedings of Presented Papers: The Parapsychological Association 33rd Annual Convention,* 168–185.

Orme-Johnson, D., et al. (1988). International Peace Project in the Middle East: The Effects of the Maharishi Technology of the Unified Field. *Journal of Conflict Resolution, 32,* 776–812.

Palmer, J. (1980). Parapsychology as a Probabilistic Science: Facing the Implications. In W. G. Roll (Ed.), *Research in Parapsychology 1979* (pp. 189–215). Metuchen, NJ: Scarecrow Press.

Protosov, V., Baron, V., Druzhkin, L., & Christyakova (1983). "Nile Elephant" Gnathonemus petersii as a Detector of External Influences, *Psi Research,2,* 31–37.

Rao, K. (1989). Meditation: Secular and Sacred—A Review and Assessment of Current Scientific Research. Presidential Address, The Indian Academy of Applied Psychology, University of Calcutta, March.

Roll, W. (1982). Changing Perspective on Life after Death. In S. Krippner (Ed.), *Advances in Parapsychological Research 3,* (pp. 147–291). New York: Plenum.

———(1988). Memory and the Long Body. *Proceedings of Presented Papers: The Parapsychological Association 31st Annual Convention,* 291–307.

Schachter, D. (1976). The Hypnagogic State: A Critical Review of the Literature. *Psychological Bulletin, 83,* 452–481.

Schmeidler, G. (1990). Is Psi a Subcognitive Process? *Journal of Parapsychology, 54,* 321–330.

Schlitz, M. J. & Honorton, C. (1992). Ganzfeld Psi Performance Within an Artistically Gifted Population. *Journal of the American Society for Psychical Research, 86,* 83–98.

Schmidt, H. (1987). The Strange Properties of Psychokinesis. *Journal of Scientific Exploration, 1,* 103–118.

Schrodt, P. (1990). A Methodological Critique of a Test of the Effects of the Maharishi Technology of the Unified Field. *Journal of Conflict Resolution, 34,* 745–755.

Solfvin, J. (1984). Mental Healing. In S. Krippner (Ed.), *Advances in Parapsychological Research 4* (pp. 31–63). Jefferson, NC: McFarland.

Stanford, R. G. (1974). Concept and Psi. In W. G. Roll, R. L. Morris, & J. D. Morris (Eds.), *Research in Parapsychology 1973* (pp. 137–162). Metuchen, NJ: Scarecrow Press.

——— (1987). Ganzfeld and Hypnotic-induction Procedures in ESP Research: Toward Understanding Their Success. In S. Krippner (Ed.), *Advances in Parapsychological Research 5* (pp. 39–70). Jefferson, NC: McFarland.

Stanford, R. G. (1990). An Experimentally Testable Model for Spontaneous Psi Events: A Review of Related Evidence and Concepts from Parapsychology and Other Sciences. In S. Krippner (Ed.), *Advances in Parapsychological Research 6* (pp. 54–167). Jefferson, NC: McFarland.

Stembridge, H. (1972). *Enhancing Creativity by Practice in Free Association While in a Hypnagogic State.* Unpublished master's thesis, University of Houston, Houston, TX.

Tart, C. T. (1972). States of Consciousness and State-specific Sciences. *Science, 176,* 1203–1210.

Underhill, E. (1911). *Mysticism: A Study in the Nature and Development of Man's Spiritual Consciousness.* London: Methuen.

Warren, C., McDonough, B., & Don, N. (1990). Single Subject Event-related Brain Potential Changes in a Psi Task. *Proceedings of Presented Papers: The Parapsychological Association 33rd Annual Convention,* 344–360.

White, R. A. (1982). Gerald Heard's Legacy to Psychical Research. *1982 Annual Proceedings of the Conference of the Academy of Religion and Psychical Research,* 56–69.

——— (1987). Meaning, Metanoia and Psi. In B. Shapin & L. Coly (Eds.), *Parapsychology, Philosophy and Religious Concepts* (pp. 167–189). New York: Parapsychology Foundation.

——— (1990). An Experience-centered Approach to Parapsychology. *Exceptional Human Experience, 8,* 7–36.

——— (1991). A New Column on Trends in Parapsychology: New views I. *ASPR Newsletter, 17* 44–46.

Witherspoon, A. (Ed.). (1951). *The College Survey of English Literature.* New York: Harcourt, Brace.

Part 6: Presidential Address

From Survival to Transcendence: Reflections on Psi as Anomalous

John Palmer (Institute for Parapsychology, 402 N. Buchanan Blvd., Durham, NC 27701–1728)

Last year at about this time I was sitting in the audience, just as you are now, listening to my predecessor and good friend Steve Braude deliver his Presidential Address to our Association. I found what he had to say very stimulating, because he talked about issues that are of great concern to me now or were of great concern to me earlier in my career. As I was listening to him, the wheels in my head started turning as I quickly tried to either formulate or resurrect how I might respond to his points, including, of course, his "reprimand du jour" about parapsychology's alleged mechanistic ways. When I discovered that I was to be Steve's successor, I realized that I had been given a golden opportunity to reply to him in the most appropriate forum possible, and I decided to take advantage of it. Will this start a trend? Will Dean Radin step up to the podium next year and comment on what you are about to hear from me? That's up to Dean, and whatever happens is fine from my point of view. I'm just honored to have the opportunity to share with you my thoughts on some important intellectual issues that parapsychology faces in the nineties.

PART I

Survival

The first issue I want to address is actually a very old one; in fact, it is one of the issues that defined our field at its inception, and it is still unresolved in most people's minds. I am referring to the question of survival of death. This section of Steve's talk reminded me of the years I spent at Ian Stevenson's Division of Parapsychology at the University of Virginia, where I gave a great deal of thought to this issue. Particularly relevant to Steve's remarks was a somewhat obscure paper I wrote there entitled "Some Recent Trends in Survival Research" that was published in *Parapsychology Review* (Palmer, 1975).

In that paper, I advocated a position similar to the one challenged by Braude in his PA Presidential Address and related papers (Braude, 1992a, 1992b, in press). The context of this position is the debate between advocates of the survival hypothesis and advocates of the so-called "super-psi" hypothesis as providing the best explanation of certain relevant empirical facts such as the phenomena of trance mediumship.[1] According to the survival hypothesis, accurate information obtained by a medium about a deceased individual is treated as evidence for that individual having survived death, especially when such information is not known to anyone present at the seance or sitting. According to the "super-psi" hypothesis, the reception by the medium of such information is more satisfactorily interpreted as simple ESP on the medium's part, with no implications for the survival of the deceased.

The typical survivalist response to this argument is that the magnitude or sensitivity of the psi in many of the better cases of mediumship is greater than in cases of psi clearly restricted to the living. Super-psi advocates reply to this argument in one of two ways. Either they deny the premise that psi in mediumship cases is superior, or more commonly, they argue that we just don't know what the limits of psi among the living are. This latter retort is essentially an appeal to parsimony: in the absence of solid knowledge about the limits of psi among the living, we should accept super-psi as the best explanation because it is simpler, in that it does not require the postulation of a new entity such as a surviving mind or personality.

The distinctions considered so far can be labeled as *quantitative*. Ostensible communications from the deceased are of the same general type as ESP among the living, although superior in scope and accuracy. However, survivalists have more recently stressed what might be called *qualitative* distinctions. The prime example of this class of phenomena are skills. Survivalists argue that skills cannot be acquired simply by receiving information about how to do something; it is also necessary to practice the skill, as in riding a bicycle (Ducasse, 1962; Gauld, 1982). Writing in defense of the super-psi hypothesis, Braude contests the claim that skills always require practice, citing as examples child prodigies and savants. For the specific case of responsive xenoglossy, which has been particularly stressed by survivalists (e.g., Stevenson, 1974, 1984), Braude argues that practice is not really necessary for learning at least some new languages; all one needs is exposure to the language. He suggests that such exposure could be provided to the unconscious by ordinary ESP.

The validity of Braude's claim is subject to empirical resolution, and I will not pretend to be enough of an expert on language acquisition to pass judgment on it. However, Braude has made an important point, namely that we cannot assume that the survivalist's argument necessarily applies to *all* skills. There may be some skills, including perhaps some language acquisition, where exposure to information is sufficient for the acquisition or development of the skill.

This caveat aside, why might the manifestation of certain skills provide better evidence for survival than does information acquisition? In the latter case, we assume that once the information is received, the capacities of the recipient's mind are quite sufficient to verbalize this information, which, of course, is how we verify that the information has been received. This is what happens in the garden variety mediumistic seance where the medium reports the message he or she receives from the ostensibly deceased communicator. In the case of skills, however, we must assume that for the skill to manifest, the information must be integrated with an aptitude that may not be present in the recipient's mind. To use a physiological metaphor, we would need to postulate not just new information coming in from the senses, but also a different neuronal network structure to process that information. To return to a dualistic metaphysics, in cases where the aptitude to demonstrate a particular skill was not present beforehand in the subject, the most straightforward solution would be that the subject's mind is replaced by a new mind. But where would this new mind come from? The most plausible solution here would seem to be that a surviving discarnate mind temporarily or permanently (as in reincarnation) possesses the subject's body. This line of reasoning suggests that, all else being equal, ostensible evidence of survival that seems to involve possession is more likely to be impressive than evidence seeming to involve telepathic communication with a discarnate entity.

From the empirical standpoint, it is necessary to demonstrate that the subject lacked the requisite aptitude before the ostensible possession took place, which in turn necessitates proving that the subject had never in the past undertaken the practice or had the experiences necessary to develop the skill. This criterion is often difficult to meet, particularly with adults, because there is no reason to investigate the aptitude until after the possession occurs, at which time it is too late. With young children this is somewhat less of a problem, provided that the skill is highly unusual for a child of a given age to possess. It is this discontinuity from the normal distribution of skills that suggests there is something in need of explanation.

Survival-related contexts. There is another factor that must be considered in evaluating the anomalous acquisition of skills as evidence for survival. In cases where survival is claimed, the skills customarily manifest in what I call a "survival-related context" (SRC). An SRC is a context in which the situation *prima facie* includes the participation of a discarnate entity. Such is obviously the case in a mediumistic seance, where the discarnate entity is presumably communicating either to or through a medium. Another example would be memories of a previous life, where the person with the memories has presumably incorporated the personality of a physically deceased individual.

The simultaneous presence of anomalous skills and an SRC provides a kind of convergent evidence on behalf of the survival hypothesis, much like specific brain wave patterns and verbal state reports provide convergent evidence for discrete altered states of consciousness (Stoyva & Kamiya, 1968). Such cases are stronger still when anomalous information acquisition, such as accurate past life memories in a reincarnation context, is added as a third kind of convergent evidence.

Convergent evidence is most impressive when the skills are present only in SRCs, because that is the circumstance under which the convergence is strongest. It is in this respect that the child prodigies and savants discussed by Braude (1994) seem particularly damaging to the survival hypothesis, because they appear to show that exceptional skills can manifest outside SRCs. In this respect, therefore, exceptional skills seem to suffer from the same weakness as anomalous information transfer as survival evidence.

But if possession is not the answer, how else can we explain why child prodigies seem to have extraordinary adult level skills at such an early age or why savants have them at all? Conventional science does not have a satisfactory answer to these questions. As long as this is true, some sort of reincarnation or possession hypothesis will still have a certain appeal in such cases, at least to anyone who is predisposed to entertain the survival hypothesis at all. Because of the absence of a strong covariance with SRCs, I certainly would not want to say at this point that child prodigies and savants provide evidence for survival. But it might be fruitful for researchers interested in the survival problem to study such persons to see if evidence of SRCs, such as previously undisclosed past life memories, might be uncovered. If they succeeded, and especially if the memories could be confirmed, the foregoing conclusion might need to be reconsidered.

Near-death experiences. I have argued that survival evidence is most impressive when there is convergence between a survival-related context and psi, and the particular kind of psi is unique to the SRC. But why does the second element have to be psi? In near-death experiences (NDEs), for example, the convergence is between an SRC and a hallucinatory experience. One advantage for psi in this role is that it can be measured objectively, whereas knowledge of the hallucinatory experience depends on the subject's verbal report, just as the SRC usually does.

The most important issue for evaluating NDEs as survival evidence, however, again comes down to how unique the hallucinations are to survival-related contexts. I think they are at least as weak as ESP in this regard, if not weaker. The capacity to visually hallucinate is very widespread in humans, particularly in dreams. The overwhelming majority of such hallucinatory experiences do not occur in SRCs. The content of the experience may seem unique to NDEs, but this can readily be interpreted as one application of the general principle that the content of hallucinations is governed by the person's concerns at the time. When a person is near death, it is hardly surprising that the hallucinations have death-related themes, and the principle of wish fulfillment can explain why the hallucinations are generally positive in tone and reflect positive themes like unification with deceased relatives or beatific deities. Qualitatively, NDEs are similar to out-of-body experiences and lucid dreams, both of which occur frequently in the population and frequently in non-SRCs. I have never understood why so many survivalists see NDEs as strong evidence for their case; in my view, NDEs are very weak evidentially in both the relative and the absolute senses.

The source-of-psi problem. Another useful way to conceptualize the survival controversy, which has received scant attention, is to treat it as a special case of the source-of-psi problem. The issue here is clearcut. Survivalists claim that in some instances the source of psi is a discarnate entity that either communicates with or possesses a mortal being. Anti-survivalists claim that the source of psi is always the mind of a mortal being.

J. B. Rhine (1960) argued that the survival problem will remain intractable until psi among the living is better understood. I do not take quite as strong a position as Rhine does. Although I agree that the kind of survival research undertaken up to this point cannot by its nature provide conclusive proof of survival, I think it can tilt the subjective probabilities for or against this hypothesis. However, I agree with Rhine to the extent of maintaining that better understanding of psi among the living could vastly improve the capacity of empirical research to solve the survival puzzle, at least to the degree that it is empirically solvable. My contribution to this discussion is to suggest that the particular problem within general psi research that needs to be addressed for this purpose is the source-of-psi problem. This problem is also highly relevant to the question of experimenter psi, which I have argued elsewhere (Palmer, 1989) is the most important research challenge facing parapsychology generally. This means that the source-of-psi problem is directly relevant to two major issues in parapsychology, the experimenter effect and the survival question.

Surprisingly little empirical research has been directed to the source-of-psi problem. The most relevant research, obviously, is telepathy research. The largest body of telepathy studies is that which compares the results in telepathy (or, more precisely, GESP) and clairvoyance conditions. The implicit model here is similar to multiple regression in that a clairvoyance condition is compared to a clairvoyance plus telepathy condition. The clairvoyance effects should cancel, meaning that any excess effect in the GESP condition can be attributed to telepathy. However, this conclusion depends on the subject being blind to the manipulation, lest expectations confound the results. This has not always been the case. Among the blind studies, results have been mixed but on balance tend to favor the telepathy hypothesis (Palmer, 1978).

Another group of studies has compared results between different telepathic agents. These studies have tended to show more success among agents known to and liked by the percipients, but again the trend is far from uniform (Palmer, 1978). If this latter methodological paradigm were to be expanded, it could provide data relevant to the survival question. Let me offer a specific suggestion, that bears some resemblance to a recent proposal by Arthur Berger (1987), that we try to identify an ideal discarnate communicator. It also bears some relation to older investigations of ostensible postmortem communications by Frederic Myers (Piddington, 1908, 1910) and to suggestions by Gertrude Schmeidler (1977).

It would be a useful project in its own right to try to find "signatures" uniquely characteristic of specific telepathic agents, along the lines of what Robert Jahn and colleagues have done with micro-PK subjects (Nelson, Dunne, & Jahn, 1984). This approach could be particularly revealing in free-response ESP paradigms, where the opportunities for identifying such signatures are relatively rich. Let's say we could find an individual advanced in age who has talent as a telepathic agent; that is to say, when a wide range of percipients are paired with this individual, they tend to score better (in the broad sense of that term) than when paired with other individuals or with no one. Let us assume further that these percipients also score better in an idiosyncratic kind of way. The possibilities for such idiosyncrasies are limitless and would have to be discovered by retrospective analyses of session transcripts.

This project could become relevant to the survival question after the individual passes away. Briefly, a group of sensitive percipients who had no prior contact with or knowledge of the deceased would be asked to participate in the same type of free-response ESP study undertaken while the deceased was still alive. A randomly selected target picture would be shown to the agent prior to his or her death. Percipients would be told that the agent would be a deceased individual but given nothing else but that person's name. The transcripts would then be evaluated blindly for the presence of the same idiosyncrasies established previously.

Although this experiment involves ESP as the convergent measure, it could provide better evidence than most mediumistic cases involving communication with a proposed discarnate entity, provided, of course, that suitably talented subjects could be found. Although super-psi could not be conclusively ruled out (it never can), the continuation after death of a reliable empirical pattern clearly linked to the agent before death would surely lend plausibility to the survival hypothesis.[2] Possession cases would still be potentially stronger, but the standards of evidentiality in such cases are much harder to meet, for the simple reason that controlled experiments are more effective in eliminating conventional alternative explanations than are spontaneous cases.

The critique of experimentation.

This brings me to Braude's "reprimand du jour," a secondary theme of which was the superiority of spontaneous case or field investigations over experiments. Braude argues that "Playing the psychic naturalist is about all we can *ever* do with psi phenomena." The whole history of parapsychology belies this assertion. The repeated demonstration that psi can be convincingly detected under controlled laboratory conditions is by itself a significant accomplishment. It is true that this evidence has not been convincing to many orthodox scientists, but this conclusion applies even more strongly to spontaneous case investigations. Moreover, our greatest strides in understanding the nature of psi have come from process-oriented experimental research. Schmeidler (1988), for example, has recently summarized a large group of studies that collectively demonstrate that psi is correlated with various psychological variables in ways that make sense from psychological theory.

This is not to say that there is no process-oriented research on spontaneous cases, however. Lousia Rhine (1965) adopted a process-oriented approach in the study of her case collections, and Roll (1976) has addressed theoretical issues in his investigations of poltergeists. We also are beginning to see some process-oriented work in the area of past life memories (Matlock, 1990). Exploring the psychological correlates of spontaneous ESP experiences, especially ones that seem likely to be genuinely paranormal, is a potentially fruitful area of psi research that we have only begun to tap.

Field studies clearly have an important role to play in understanding psi. By definition they are more directly relevant to psi as it occurs in the real world than are experiments. A good case study can often be more convincing that a mediocre experiment. However, on the average, experiments are superior because of their greater capacity to control against alternative interpretations. This capacity is central to what scientific investigation is all about. The essence of scientific understanding is to identify all the possible explanations of an observation and rule out all the possibilities but one. Experiments are generally better equipped than field studies to meet this challenge. Nonetheless, the ideal outcome is a convergence of evidence from both experimental and field studies, the former providing the control and the latter providing the ecological realism. Field studies can also provide ideas or hypotheses for more incisive experimental research.

Another distinguished member of our Association, Rhea White (1990), takes a more radical position than Braude does. Echoing Susan Blackmore (1988), White argues that all attempts to establish objectively that psi is "real," whether by experiments or by field studies, have been a failure and should be abandoned, although unlike Blackmore she leaves open the possibility that such attempts might be resumed at a later date. As the replacement, White appeals for a "person-centered" or phenomenological study of psi experiences for the purpose of understanding their subjective meaning to the experiment. We might say that the quest for an understanding of objective psi is to be replaced by a quest for an understanding of subjective psi. The latter is certainly a worthwhile goal, and I even suspect that research directed toward this goal would be more fruitful in a short span of time than the search for an understanding of objective psi has been. A better phenomenological understanding of psi experiences could also benefit traditional psi research. For instance, if we could better understand why psi experiences are meaningful to people, we could use this knowledge to create more psi-conducive contexts in the laboratory. I have felt for years that depth psychology has a potentially important role to play in the understanding of both subjective and objective psi, and I am beginning to incorporate it in my own research.

What troubles me about proposals like those of White and Blackmore is their exclusivity. Our progress in gaining an understanding of objective psi has indeed been frustratingly slow and that makes it an easy target for attack. But the goal remains a legitimate and important one, and we should not abandon it, or even postpone it, just because it is hard to attain. Let's not emulate the proverbial behavioristic drunk who looks under a lamp post one night for the keys he dropped in the gutter because the light under the lamp post is better. If we were to accept White's philosophy in science generally, we might never find a cure for illnesses like breast cancer or the common cold. We must expect progress in such a difficult area as parapsychology to be slow at first, but slow does not mean negligible or nonexistent. White herself acknowledges that some progress has been made in understanding objective psi, as, for example, regarding the experimenter effect, and we can expect progress to accelerate as we focus more and more on process-oriented pursuits. Although phenomenological research has a role to play in this process, it does not provide an adequate methodological substitute for attaining the kinds of understanding of psi that parapsychologists have traditionally sought.

Should the phenomenological study of psi experiences even be considered a part of parapsychology? I used to believe that it would be better to classify it as a subtopic of cognitive or transpersonal psychology. I changed my mind when I began thinking of this issue in relation to my proposition that parapsychology should be defined by its subject matter rather than by its theoretical or quasi-theoretical constructs (Palmer, 1988); parapsychology's subject matter, of course, is psi events, and these events often have phenomenological components. I now believe that any form of inquiry about psi events should be welcomed into parapsychology's domain.

The stage is now set for a discussion of the essence of Braude's "reprimand." He phrases his objection as parapsychology's infatuation with mechanistic thinking, although his examples suggest that what he is complaining about can be more incisively labeled as reductionism. In particular, Braude cites the alleged deference of parapsychology to the methods and concepts of Physics. Such an approach is reductionistic to the extent that it attempts to account for psi in terms of things like fields and subatomic particles. Another example of a reductionistic approach is brain research, including disciplines such as Psychophysiology and Psychopharmacology. Braude argues that such reductionistic methods are not appropriate to the study of psi.

To begin with, Braude overrates the degree to which parapsychology is reductionistic or mechanistic, at least as I use these terms. Most of the psychologically-oriented theory and research in parapsychology use mentalistic constructs that have no direct, and in some cases not even an indirect, relation to brain structures or brain chemistry. Much of it, such as Rex Stanford's revised PMIR and conformance behavior models (Stanford, 1978, 1990), can best be labeled as teleological. Probably the most reductionistic of parapsychological research programs are the micro-PK studies and surrounding theory. But even in this domain the reductionism is far from complete. Helmut Schmidt (1975) treats psi as goal-oriented (Stanford, 1990), which is another example of teleological thinking. The observational theorists use consciousness as a causal construct (e. g., Walker, 1974), and consciousness is even more prevalent in the model of Jahn and Dunne (1987). The pragmatic information model of Walter von Lucadou (1984, 1988) has a very holistic emphasis derived from general systems theory, and such an emphasis also characterizes Roll's (1982) psychic structure theory. In most of these conceptions, the psi process per se is treated as primitive and unanalyzable.

Moreover, I see much positive value in some reductionistic research programs. A particularly neglected area in psi research is the possible effect of brain chemistry on psi. There are tantalizing anecdotes of hormonal imbalances or concussions being associated with strong psi abilities. With the possible exception of Bill Delmore (Kanthamani & Kelly, 1974), the two highest scoring card-guessing subjects in terms of effect size chronicled in the *Journal of Parapsychology* both had thyroid imbalances (Drake, 1939; Riess, 1937, 1938). Explorations in these areas could have much to tell us about how brain processes mediate psi, and they might also guide us in selecting gifted subjects or in enhancing psi ability. The point here, as Braude himself (1986, 1987) has noted on numerous occasions, including his Presidential Address (Braude, 1994), is that the same phenomena can be profitably understood from a variety of conceptual frameworks. Some of these frameworks will undoubtedly be reductionistic.

A related charge that Braude makes in his Address is that too many experimental parapsychologists accept the "small is beautiful" assumption. In this case the charge is clearly accurate; laboratory investigators do by and large study smaller magnitude events than do field investigators. This is particularly true in the case of PK. However, I believe the "small is beautiful" assumption can be defended. Experimental parapsychologists do not favor small effects because they are small, or because we feel a need to kowtow to Physics, but because on the average small effects have proven to be easier to control than large effects, not to mention the fact that it is easier to find subjects capable of demonstrating them. Because small effects are less dramatic, they also are less likely to attract charlatans and publicity-seeking magicians, along with the three ring circus atmosphere that inevitably accompanies them. But the main point is that the goal of science is conceptual understanding, and for that purpose controllability is more important than magnitude.

There is one more aspect of Braude's argument that I have yet to address, and I believe it is the most powerful aspect. It brings us back to the source-of-psi problem. Citing the seminal writings of Jule Eisenbud on this topic (1982), Braude suggests that conventional experimental methods are inappropriate to psi because the nature of psi precludes the application of standard experimental controls. His references make clear that the main thing that cannot be controlled is the source of psi. How, for example, can our research tell us anything reliable about a subject's psi ability when for all we know we are really measuring the experimenter's psi ability? I addressed this problem in a previous paper (Palmer, 1989), so let me just summarize my position here. If Braude is right that standard controls cannot be applied in psi research, then many conceptual questions about the nature of psi will never be answered. Ironically, this statement applies not only to laboratory investigations but also to the field studies of interest to Braude. As we saw in the discussion of survival research, source-of-psi questions often enter into our attempts to interpret what is going on outside as well as inside the lab.

However, I am not as pessimistic as Braude. I believe that standard experimental methods creatively applied can deal with the source-of-psi problem, as illustrated by my suggested research project to address the survival hypothesis. Moreover, some lawfulness has been shown in psi data, as for example the sheep-goat effect (Palmer, 1971). Such data support the notion that the world of psi is not as chaotic as the writings of Eisenbud and Braude might suggest it is, and this offers the prospect of further progress in the experimental arena.

PART II

Psi as Anomalous

Several years ago I authored a series of papers advancing suggestions for the reconceptualization of the psi controversy (Palmer, 1986, 1987, 1988). In a nutshell, I proposed that psi be redefined as a descriptive, atheoretical term and that the fundamental question anchoring the controversy be changed from "Does psi (in the theoretical sense) exist?" to "How can psi (in the descriptive sense) be best explained?" The question "Does psi exist?" has two unfortunate repercussions. First, it encourages proof-oriented rather than process-oriented research, whereas I strongly believe the latter is what we need to advance the field. Second, it has the unfortunate rhetorical consequence of allowing a failure to prove that "psi exists" to be interpreted as justifying the conclusion that "psi does not exist," or that the phenomena have adequate conventional interpretations.

In contrast, the preferred question, "How can psi be best explained?" defines psi neutrally as a phenomenon in need of explanation and thus encourages process-oriented research. Second, it allows for and encourages a distinction between inability to confirm paranormal explanations and success in confirming conventional explanations. In this connection, I noted that there are three alternative answers to the question "How can psi be best explained?": (1) Psi can be best explained by paranormal constructs; (2) psi can be best explained by conventional scientific constructs; and (3) psi at present cannot be adequately explained by either paranormal or conventional constructs. I opted for the third of these answers as the correct one for this point in time, and I associated this position with the label "anomalous." Specifically, I maintained that psi should be advanced by the parapsychological community as a genuine scientific anomaly for which science has as yet no adequate explanation.

I saw two principal advantages to this position. First, I thought it was the most intellectually valid way to describe the current state of the evidence. In the absence of any explanatory theory of psi that has sufficient empirical support to justify broad acceptance, there is no justification to conclude that the best explanations of our phenomena must be paranormal. We simply don't know what the best kind of explanation will prove to be, and we may not have considered all the possible alternatives. Second, I thought the anomaly position, if handled properly, would provide advantages in the rhetorical battles with our critics. The claim that our phenomena are anomalous is indeed more modest than the claim that they are paranormal, but it is also easier to defend. It is relatively easy for critics to argue that we have not proven that our phenomena require paranormal explanations, but much more difficult for them to argue that the phenomena can be explained satisfactorily by conventional science. When critics are forced to accept their fair share of the burden of proof, they must allow their own explanations of the phenomena to undergo critical scrutiny. Because these explanations are often implausible and blatantly ad hoc, they usually don't fare well under this scrutiny. Hyman, in his parapsychology chapter of the infamous *National Research Council Report* on techniques for enhancing human performance (Druckman & Swets, 1988), had to admit that there were no plausible conventional explanations for some of the better psi evidence, and this became a cornerstone of our response to the *Report* (Palmer, Honorton & Utts, 1988).

Attacks by critics. I have been curious how critics would respond to my proposals. Would they accept the notion that the 'psi as anomalous' claim is valid, thereby joining with conservative parapsychologists to form a powerful intellectual alliance that could define the scientifically acceptable position on psi? Or would they claim that psi has not even been established as anomalous, thereby radicalizing their own previous position? I suppose it should not be too surprising that what responses there have been have fallen into the latter category.

One of these responses was by James Alcock (1990). As I noted in a review of his recent book *Science and Supernature* (Palmer, 1991), Alcock "sometimes writes as if *anomalous* is just a less offensive synonym for *paranormal* and then uses the standard arguments against the old claim to undermine the new one. For instance, he writes that 'even if paranormal were to be defined only in terms of anomaly, this would still lead to a dualism of some sort because of its independence from considerations of time and space (p. 18)' " (p. 88). The point of the term anomalous, however, is that it is theoretically and metaphysically neutral and thus does not require that the phenomena be independent of space and time or that we adopt a dualistic ontology.

The most substantial response to my thesis has been by Ray Hyman (Druckman & Swets, 199–200), who, unlike Alcock, understands the distinction between anomalous and paranormal. Hyman makes two relevant points. First, he notes that psi cannot be construed as a single anomaly because the phenomena may have a number of different explanations. Second, he notes that anomalies in other sciences are more reliable and precisely defined mathematically than is psi. My response to both points is so what. Most parapsychologists already acknowledge that different psi phenomena may have different explanations when we distinguish between ESP and PK. Furthermore, we all would agree that some ostensible psychic phenomena have mundane conventional explanations. The key point is that because anomalous is an atheoretical term, distinctions having to do with the correct explanations of the phenomena are irrelevant to their classification as anomalous. One thing that unites psi phenomena in this category is that they lack any adequate explanation at present but are prima facie paranormal. Regarding Hyman's second criticism, the fact that psi phenomena are relatively unreliable and poorly quantified is also irrelevant to their being labeled as anomalous. Puzzles do not cease to be puzzles because they are hard to pin down. In short, Hyman's attempt to refute the claim that psi is a genuine scientific anomaly is extremely weak.

Attacks by parapsychologists. The anomaly construct has been challenged on much different grounds by some parapsychologists. The common theme is that by referring to psi as anomalous we trivialize it. In a guest editorial in the *Journal of the American Society for Psychical Research,* Theodore Rockwell (1989) wrote "Those who would define psi phenomena as mere anomalies would by this action reduce parapsychologists to chroniclers of Forteana. From being a systematic attempt to explain certain widely experienced human phenomena, parapsychology would become merely a matter of cataloguing curiosities" (p. 201). In a book review in that same journal, Emily Cook (1991) wrote, "There are frequent debates in parapsychology today about the definition and scope of the field, with many parapsychologists defining it as the study of 'anomalous' or 'unexplained' experiences. Not surprisingly, as a result there are frequent criticisms that the field is defined negatively, that it has no theoretical basis, and that it exists only because scientists have so far failed to explain the phenomena and that it shrinks with each 'successful' explanation" (p. 189). Finally, Rhea White (1990) stated that "If we hope to understand the phenomena of parapsychology, it is wrong to call them anomalous because that places them in a context devoid of meaning" (p. 2).

Comments like these reflect a misunderstanding of what the term anomalous is meant to connote in parapsychology or the role it is supposed to play. To begin with, neither I nor anyone else I know of who uses this term is saying that psi should be indiscriminately lumped together with other mysteries of nature into a category defined solely by the "mysteriousness" of its content. Anomalous in parapsychology is always meant to be secondary to the term psi, which restricts our anomalies to the relatively small class of phenomena that parapsychologists study. There is some debate within parapsychology about the scope of the term psi, but most of us agree that it restricts our inquiry to those events that are prima facie examples of communication processes between minds (human or possibly subhuman; incarnate or possibly discarnate) and their external environment. Psi remains the fundamental label for our phenomena, and in most discourse it can be used alone without any explicit reference to the term anomalous.

Second, psi is negatively defined only if it is considered equivalent to paranormal, which is how it is understood in the traditional conceptual framework I am challenging. If, as I have suggested, psi is used descriptively to represent the actual phenomena that are thought to require a paranormal explanation, the definition becomes at least partially positive: that is, psi refers to a subclass of anomalous phenomena with specific properties that characterize them as communication-like interactions.

Third, although the term anomalous implies that there is as yet no adequate explanation for these phenomena, it does not preclude such an explanation, nor does it in any way discourage the quest for this explanation or deny that viable candidates are being systematically pursued. It merely says that the goal has not yet been reached. A noncommittal stance on the paranormality of psi in no way precludes process-oriented research based on the reasonable working hypothesis that any psi uncovered in a properly controlled investigation is paranormally caused.

But all unexplained facts of nature are not called anomalous. Why do we need this term in parapsychology? The reason is that psi appears to be incompatible (or at least incommensurate) with widely accepted basic laws of nature. Rockwell thinks we make too much of this fact, and I agree. But like it or not, there are many other people who think psi's incommensurability with other physical laws is very important. If our vocabulary is to be realistic, it must incorporate this fact and that is why we need the term anomalous. We need a term like paranormal, which Rockwell and I also dislike, for the same reason. We must have a generic label for those alternative theories that do not mesh with our current ones, and unless we resort to a new term like my ill-fated omega (Palmer, 1988), paranormal is it. If we were willing to restrict ourselves to those potential explanations of our phenomena that are consistent with current scientific laws, as our critics would have us do, then we would not need terms like anomalous or paranormal. But we aren't willing to do that, so these terms are necessary in certain contexts.

I suspect that what underlies much of the antagonism toward the anomaly concept is that it concedes too much to our critics. But this concession is much less apocryphal than it might appear on the surface. First, the establishment of a genuine scientific anomaly is itself an important accomplishment. As Thomas Kuhn (1970) has pointed out, such anomalies have often been the springboard to important scientific advances. What more would be accomplished if in addition we could justifiably claim that paranormality has been established? We could rule out a class of explanations that we find relatively uninteresting theoretically, but where would that really get us? I submit, not very far. When we say that psi is paranormal, we are saying what it is *not,* but the significance of our phenomena must be derived from an understanding of what psi *is*. To say what psi is, to put some meat on the bones, we need a positive theory of psi that directly explains it. But even most of those who would insist that psi is paranormal would agree that we have yet to establish a comprehensive explanatory theory of psi that demands broad acceptance. From the applied perspective, the paranormality of psi has no relevance at all, because the issue there is what can be achieved, not what the explanation is. Finally, we can point out that when critics are unable to propose a plausible conventional interpretation of a given psi event, the likelihood that the event was paranormal is large, even if not conclusively demonstrated.

Process-oriented research. One important implication of the anomaly position is that the challenge facing our field is not to demonstrate psi but to understand it. The focus must be on theory development and testing and on process-oriented research. This applies to field studies as well as to laboratory experiments. In the experimental context this requires that we find ways to make psi data more reliable, because the subtle correlational effects that must be uncovered to sustain incisive theoretical interpretations of psi effects are likely to be obscured by the large components of error variance in the psi test scores we customarily work with now. These attempts to increase reliability are likely to be most successful if they are conceptually driven. Labelling psi as anomalous reminds us that its demonstration is only one step toward the important prize, and a step we have already achieved. Thus, if properly understood, the anomaly concept actually favors the kind of process-oriented research that those parapsychologists who criticize the term also seem to be advocating.

This same philosophy applies to psi applications. Increasing attention has been paid in recent years to the success of psi applications in the real world, particularly those involving police work (Lyons & Truzzi, 1991; but see also Hyman, 1992). More systematic research is still needed, but it is likely that in at least a handful of cases input from psychics may have been of positive benefit. However, the very newsworthiness of these possibly successful applications highlights how rare they are and how relatively small a role they play in solving the broader problems to which they are addressed. It is clear that the abilities of even our best psychic talents are only sufficient to make a small dent in the ubiquitous challenges of curing diseases and solving crimes. If psi is to make any real difference in the world, the level of functioning of its practitioners, impressive though it may sometimes be in a research context, must be substantially enhanced. If this is to happen, however, it will require the kind of understanding of the psi process that is the goal of basic, process-oriented research. Paradoxical as it may sound, basic research is as important and irreplaceable in the applied context as it is in the theoretical context.

Gaining Public Support

A final criticism that might be levelled at the anomaly concept concerns the mundane matter of salesmanship. We are all painfully aware of the precarious financial situation in which parapsychology now finds itself. It seems that almost all our institutions have either closed their doors or are in varying degrees of financial difficulty. It is vital for the field's continued progress, if not its very survival, that new sources of funding be developed.

I agree that a conservative term like anomalous is unlikely to inspire a potential donor, but no one is proposing that we use this term as a cornerstone in our appeal for funds. On the other hand, I do believe that for reasons of intellectual honesty we must communicate the status of the evidence that the term anomalous implies: that we have exciting findings, but we really don't know yet what they mean. I think that some modesty in our claims might actually be beneficial to parapsychology's image. I've never done a formal survey on this, but I have the distinct impression that in science generally, the less exaggerated the claim, the higher the status of the individual making it or the status of the specialty he or she represents. In the medical field, in particular, it is customary to hear great caution expressed that such and such a finding is not yet a breakthrough and that more research, replication, etc., is needed. In more "fringe" areas, the claims are sometimes more exaggerated, and the claimants often come across like used car salesmen. I've seen examples of both kinds of claims in parapsychology, and I believe the more modest claims are more impressive because they are more credible.

Another advantage of more modest claims is that they highlight the fact that more research is needed. When I was working in California, I sometimes heard New Age aficionados argue that psi had already been proven scientifically so there is no longer any need for research. A rhetorical approach that emphasizes what still needs to be learned about psi can help overcome this kind of complacent attitude, which I doubt is restricted to New Agers. At the same time, of course, we must persuade these individuals that our research up to this point has made progress, so they will feel we have the ability to advance further. It is for this reason that our impressive meta-analyses (Utts, 1991) and other evidence that we have uncovered a genuine scientific anomaly must be successfully marketed.

An equally important but, ironically, a probably more difficult task is to persuade people that the scientific method is our best hope for achieving an understanding of psi. Practicing parapsychologists take this for granted, but many of the people we need to persuade do not. Parapsychologists are an anomaly of sorts in that we are exceptions to a trend I see in the population at large; namely, a negative correlation between an interest in psi and an appreciation of science. Ironically, this is where we need to mimic CSICOP by serving as cheerleaders for the methods and accomplishments of traditional science. To be sure, we object to the rigid materialism of traditional science and the way it has treated parapsychology, but the fundamental methods and epistemology of science are part of the very essence of our field. Without them, parapsychology would not exist. We must be positive advocates of them and what they have accomplished in other domains if we hope to justify our existence.

The implications of psi. But all this is not enough to really inspire the public to support parapsychology. A key ingredient is missing, and that ingredient, I think, is a no-holds-barred discussion of the possible implications of psi. It is not an overstatement to say that these potential implications are monumental. Part of J. B. Rhine's success in gaining public support for parapsychology was that he was not afraid to openly discuss its implications in his popular books on the subject. But today we see less of this, and I think that is a major reason why we have attracted so little tangible public support. It is also bad for our own morale. I suspect that many of us get so caught up in the nuts and bolts of our specific research projects and other duties that we forget about the idealism that brought us into parapsychology in the first place. We are engaged in a very important enterprise for which many of us have made great sacrifices, and we should take time to pat ourselves on the back for choosing this path. Most importantly, we should continually remind ourselves what this struggle is all about and draw energy from that realization in our day-to-day existence.

But wouldn't talk about the potential implications of psi risk alienating members of the traditional scientific community whose support we also need? It is fashionable in neo-positivistic circles to look down upon such discussions as unjustified breast-beating, despite the fact that they are customarily required in grant applications to federal agencies. This criticism nonetheless has some validity, but only insofar as claims about the implications of a particular finding go beyond the data. However, this caveat does not preclude a discussion of what the implications *might be,* as distinguished from what they "*are.*" Indeed, such cautious discussion of the potential implications of a research program can be considered a morally necessary way to provide those who would support the research program with an understanding of what their support might lead to down the road.

I would predict that even a very cautious public discussion of the implications of psi would enrage those scientists for whom the current scientific world view functions as a kind of quasi-religion, but these people are already hostile to our field and will continue to be so whatever we do. I think more important irritants among scientists generally are the exaggerated claims sometimes made for psi, sensationalism in the media, and the identification of the field with occultism. These are problems that we are already aware of and dealing with as best we can. If we can place a discussion of the possible implications of psi in a context that projects a responsible attitude toward these other problems, the discussion might actually have a positive impact on many scientists. In addition, it would help for us to make a point that also has been stressed by Braude (1986), namely that confirmation of a paranormal theory of psi would not mean that we must abandon existing laws of materialistic science, only redefine their boundaries. The opposite claim is often stressed by our critics, but it is silly to think that Newtonian mechanics, for example, would suddenly become invalid the day some scientific committee declares that paranormal processes are real.

So what are the implications of psi? They fall into two main categories, the applied and the conceptual. Let us first consider the potential applications of psi. I sometimes find it amusing to see these potential applications restricted to discrete subject areas like healing and crime solving. The fact is that we are talking about the potential for a marked and broad increase in human capacity to interact with our external environment, both to acquire information about it (ESP) and to change it (PK). If the kinds of phenomena already demonstrated in our research could be maximized and brought under better control, the possibilities are endless. If we can psychokinetically increase the rate at which a wound heals (Wirth, 1990), is it all that much of an extrapolation to suggest that someday we might be able to psychokinetically dissolve a brain tumor? If we can use precognition to learn the future behavior of a computer (Radin & Nelson, 1989), is it too much of an extrapolation to suggest that someday we might use it to forecast a devastating earthquake?

A serious discussion of such applications of course raises ethical issues. Any increase in human capacity, whether it be psychic powers or the benefits of education, can be used for either good or evil purposes. I tend to be an optimist in this regard. When one looks at the history of civilization, which embodies the increases in the capacities of human beings that help define our civilization, man's inhumanity to man has on balance decreased, despite the horror stories we still read about in the newspapers. Civilization has advanced in spite of its increased capacity for self-destruction. I think this is a function of education in the broad liberal sense. When I look at the world's great tyrants, people like Hitler and Saddam Hussein, I see people who are clever but not very well educated in this broad sense. I think that the more complete education that psychic means of information acquisition could provide on a grand scale would include an appreciation of humane moral values that would render the use of psi for evil purposes much less likely than it seems today. There inevitably would remain some people who would use their advanced psi capacities for evil purposes, but these capacities would be more than offset by the increased capacities of those with more virtuous motives to counter the evil.

This optimistic scenario is predicated on the assumption that human nature is basically good, and that evil motives arise primarily from ignorance. It also assumes that psychic abilities would be broadly based and not the sole province of some treacherous elite. These assumptions and the scenario that follows from them are, of course, debatable, but the fact remains that whatever stance one takes, there is much at stake in the potential fulfillment of psi research. If we choose to pursue these avenues, there are many who will be prepared to provide tangible support for our quest.

The potential conceptual implications of psi are also dramatic, but the ethical issues they raise, although existent, are not nearly so profound as from the applied perspective. Moreover, the realization of these conceptual implications would not require the same strength of effect that will be needed for impactful application. The conceptual implications are more difficult to foresee, however, because they depend heavily on what theory of psi we ultimately adopt.

If the conventionalists are right, the implications of psi will primarily involve methodology in behavioral science. As Hyman and Honorton (1986) have pointed out, a "failed" psi research program could give us a better understanding of how subtle artifacts might compromise research findings in conventional scientific fields, particularly psychology. This, of course, is not the kind of implication most of us expect or desire, but it does illustrate that even in the worst case scenario our labors would not have gone entirely for naught.

If psi, on the other hand, proves to have a paranormal explanation of some sort, the implications might be more profound. Let me just list a few of the more likely ones as I see them, beginning with scientific explanations. Our fundamental theories of Physics would clearly require a mental component, such as we have already seen in the introduction of consciousness into quantum mechanics by some theorists (Walker, 1974). However, these mental constructs would need to be more refined to allow for probabilistic mathematical laws relating changes in mental state to the "behavior" of matter. The main point is that mind would be seen as the prime mover in nature.

We also must consider the philosophical and cultural implications if such paranormal theories of psi were to be confirmed. Here I am talking not about scientifically verified theories, but about the impact of these theories on beliefs that are not strictly amenable to scientific proof. Ontologically, the Physics described above would seem to favor dualism (Beloff, 1964) or Whitehead's panexperientialism (Griffin, 1993), although materialism might survive if it could be shown that these mental effects require mortal beings. This of course brings us back to the survival question. Perhaps the most important intellectual implication of psi research will be its capacity to answer the question of whether our consciousness can transcend our mortality.

Any paranormal theory of psi would lend credibility to religions which affirm paranormal occurrences. On the other hand, some more fundamentalist religions might be undermined because events that were theretofore considered to be examples of supernatural or divine intervention would have to be reinterpreted as conforming to general scientific laws. However, religion in the broader sense would be enhanced, I think, because the notion of God is more compatible with a fundamentally mental universe than a fundamentally material one.

Finally, I believe that psi has the potential of lending credibility to the world view of transpersonal Psychology, particularly with respect to the notion of our interconnectedness. My bet is that the theory of psi we ultimately adopt will have as its basis a kind of collective unconscious, a universal mind that we all share. Ordinarily, we only have access to those contents of the universal mind that support our sense of individual identity. Psi occurs when we can break out of this identity long enough to access other information in the universal mind. I believe that the cultural consequence of such a metatheory, and particularly experiences that would arise from its application, would be a greater empathy with and respect for our fellow human beings. This is part of why my attitude about the moral implications of advanced psychic powers tends to be positive.

There is little if anything I have said in this speculative exercise that has not been said by others before me. My purpose is not so much to provide new insights as to remind us of exactly what is at stake in our research. Proper validation of a paranormal theory of psi would, like other major scientific theories before it, have enormous consequences for how we perceive ourselves and the world around us. If these insights could be applied for potential use, the consequences would be even greater. The primary accomplishment of our research to date has been to make the possibility of these consequences sufficiently great that they merit serious discussion. This is not an accomplishment that should be underestimated or undervalued.

The problem is that these potential consequences are not real to people. Surveys tell us that lots of people believe in psi, and television ratings tell us that lots of people are interested in psi. But I don't think many people really appreciate or have given much thought to the implications of psi, what could be at stake for them both individually and collectively. I submit that this is the major reason why this interest in parapsychology is passive rather than active, why it is something that is fun for party conversation and bedtime reading rather than something to get involved with and promote. Only if we can sensitize the public to these possible implications (even if it leads to some controversy), and only if we can then convince them that the scientific method is the way to find out if these potential implications are real, will we parapsychologists receive the kind of public support we need and deserve.

NOTES

1. A third possibility is represented by the "psi structures" theory of William Roll (1982). According to this view, we do not survive as intact personalities but are represented as dispositions attached to objects or persons that we had prior interactions with.
2. A specific non-survivalist counterhypothesis for this experiment that involves retrocognition was suggested to me after my oral presentation by Dr. Christopher Cooper. I confess that I had not considered a retrocognition interpretation of my experiment, perhaps because parapsychologists hardly ever think about retrocognition at all. The hypothesis is simply that the agent could have undertaken the communication before death (perhaps unconsciously), with the information stored unconsciously in the percipient's mind until the free-response ESP trials.

   I see no way to rule out this version of the super-psi hypothesis, although it might be weakened by introducing a long delay (say, six months) between the agent's death and the experimental ESP trials. The plausibility of this move would be increased if it could be shown that retrocognition among the living tends to atrophy as the time interval increases. This point illustrates how more knowledge about psi in general can help in solving the survival puzzle. I could further reduce the plausibility of the retrocognition hypothesis by not selecting the percipients until just before the ESP trials are to commence.

REFERENCES

Alcock, J. E. (1990). *Science and Supernature: A Critical Appraisal of Parapsychology*. Buffalo: Prometheus.

Beloff, J. (1964). *The Existence of Mind*. New York: Citadel.

Berger, A. S. (1987). *Aristocracy of the Dead: New Findings in Postmortem Survival*. Jefferson, NC: McFarland.

Blackmore, S. J. (1988). Do We Need a New Psychical Research? *Journal of the Society for Psychical Research, 55,* 49–59.

Braude, S. E. (1986). *The Limits of Influence: Psychokinesis and the Philosophy of Science*. New York: Routledge & Kegan Paul.

——— (1987). How To Dismiss Evidence Without Really Trying. *Behavioral and Brain Sciences, 10,* 573–574.

——— (1992a). Reply to Stevenson. *Journal of Scientific Exploration, 6,* 151–155.

——— (1992b). Survival or Super-psi? *Journal of Scientific Exploration, 6,* 127–144.

——— (1994). Psi and the Nature of Abilities. *Research in Parapsychology 1991,* 193–219.

Cook, E. W. (1991). Psychology and Parapsychology: Some comments on Paul Kline's *Psychology Exposed, or the Emperor's New Clothes*. *Journal of the American Society for Psychical Research, 85,* 183–192.

Drake, R. M. (1938). An Unusual Case of Extra-Sensory Perception. *Journal of Parapsychology, 2,* 184–198.

Druckman, D. & Swets, J. A. (Eds.). (1988). *Enhancing Human Performance. Issues, Theories, and Techniques*. Washington, DC: National Academy Press.

Ducasse, C. J. (1962). What Would Constitute Conclusive Evidence of Survival After Death? *Journal of the Society for Psychical Research, 41,* 401–406.

Eisenbud, J. (1982). *Paranormal Foreknowledge: Problems and Perplexities*. New York: Human Sciences Press.

Gauld, A. (1982). *Mediumship and Survival*. London: Heinemann.

Griffin, D. R. (1993). Parapsychology and Philosophy: A Whiteheadian Postmodern Perspective. *Journal of the American Society for Psychical Research, 87,* 218–288.

Hyman, R. (1992). [Review of *The Blue Sense.*] *Journal of Parapsychology, 55,* 399–408.

——— & Honorton, C. (1986). A Joint Communique. The Psi Ganzfeld Controversy. *Journal of Parapsychology, 50,* 351–364.

Jahn, R. G. & Dunne, B. J. (1987). *Margins of Reality: The Role of Consciousness in the Physical World.* New York: Harcourt Brace Jovanovich.

Kanthamani, H. & Kelly, E. F. (1974). Card Experiments with a Special Subject. I. Single-card Clairvoyance. *Journal of Parapsychology, 38,* 16–26.

Kuhn, T. S. (1970). *The Structure of Scientific Revolutions.* (2nd ed.). Chicago: University of Chicago Press.

Lucadou, W. von (1984). What is wrong with the definition of psi? *European Journal of Parapsychology, 5,* 261–283.

——— (1988). The Model of Pragmatic Information (MPI). In D. H. Weiner and R. L. Morris (Eds.), *Research in Parapsychology 1987* (pp. 18–22). Metuchen, NJ: Scarcrow Press.

Lyons, A. & Truzzi, M. (1991). *The Blue Sense: Psychic Detectives and Crime.* New York: Mysterious Press.

Matlock, J. G. (1990). Past Life Memory Case Studies. In S. Krippner (Ed.), *Advances in Parapsychological Research: Vol. 6* (pp. 184–267). Jefferson, NC: McFarland.

Nelson, R. D., Dunne, B. J., & Jahn, R. G. (1984). An REG Experiment with Large Data Base Capability, III: Operator Related Anomalies. (Technical Note, PEAR 84003). Princeton Engineering Anomalies Research, Princeton University.

Palmer, J. (1971). Scoring in ESP Tests as a Function of Belief in ESP. Part I: The Sheep-Goat Effect. *Journal of the American Society for Psychical Research, 65,* 373–408.

——— (1975). Some Recent Trends in Survival Research. *Parapsychology Review, 6*(3), 15–17.

———(1978). Extrasensory Perception: Research Findings. In S. Krippner (Ed.), *Advances in Parapsychological Research: Vol. 2.* New York: Plenum.

——— (1986). Progressive Skepticism: A Critical Approach to the Psi Controversy. *Journal of Parapsychology, 50,* 29–42.

——— (1987). Have We Established Psi? *Journal of the American Society for Psychical Research, 81,* 111–123.

——— (1988). Conceptualizing the Psi Controversy. *Parapsychology Review, 19*(1), 1–5.

——— (1989). Confronting the Experimenter Effect. *Parapsychology Review, 20*(4), 1–4; *20*(5), 1–5.

——— (1991). [Review of *Science and Supernature*.] *Journal of Parapsychology, 55,* 84–89.

———, Honorton, C., & Utts, J. (1988). *Reply to the National Research Council Study on Parapsychology*. Research Triangle Park, NC: Parapsychological Association.

Piddington, J. G. (1908). A Series of Concordant Automatisms. *Proceedings of the Society for Psychical Research, 22,* 59–77.

——— (1910). Three Incidents from the Sittings: Lethe; the Sibyl; the Horace Ode Question. *Proceedings of the Society for Psychical Research, 24,* 86–104.

Radin, D. I. & Nelson, R. D. (1989). Evidence for Consciousness-related Anomalies in Random Physical Systems. *Foundations of Physics, 19,* 1499–1514.

Rhine, J. B. (1960). Incorporeal Personal Agency: The Prospect of a Scientific Solution. *Journal of Parapsychology, 24,* 279–309.

Rhine, L. E. (1965). *Hidden Channels of the Mind.* New York: William Morrow.

Riess, B. F. (1937). A Case of High Scores in Card Guessing at a Distance. *Journal of Parapsychology, 1,* 260–263.

——— (1939). Further Data from a Case of High Scores in Card-guessing. *Journal of Parapsychology, 3,* 79–84.

Rockwell, T. (1989). Guest Editorial: Psi Is Not a Mere Anomaly. *Journal of the American Society for Psychical Research, 83,* 201–203.

Roll, W. G. (1976). *The Poltergeist.* Metuchen, NJ: Scarecrow Press.

——— (1982). The Changing Perspective on Life After Death. In S. Krippner (Ed.), *Advances in Parapsychological Research: Vol. 4* (pp. 147–291). New York: Plenum.

Schmeidler, G. R. (1977). Looking Ahead: A Method for Research. *Theta, 5*(1), 2–6.

——— (1988). *Parapsychology and Psychology: Matches and Mismatches*. Jefferson, NC: McFarland.

Schmidt, H. (1975). Toward a Mathematical Theory of Psi. *Jour-*

nal of the American Society for Psychical Research, 69, 301–319.

Stanford, R. G. (1978). Toward Reinterpreting Psi Events. *Journal of the American Society for Psychical Research, 72,* 197–214.

——— (1990). An Experimentally Testable Model for Spontaneous Psi Events: A Review of Related Evidence and Concepts from Parapsychology and Other Sciences. In S. Krippner (Ed.), *Advances in Parapsychological Research: Vol. 6* (pp. 54–167). Jefferson, NC: McFarland

Stevenson, I. (1974). *Xenoglossy: A Review and Report of a Case.* Charlottesville: University Press of Virginia.

———(1984). *Unlearned Language: New Studies in Xenoglossy.* Charlottesville: University Press of Virginia.

Stoyva, J. & Kamiya, J. (1968). Electrophysiological Studies of Dreaming as the Prototype of a New Strategy in the Study of Consciousness. *Psychological Review, 75,* 192–205.

Utts, J. (1991). Replication and Meta-analysis in Parapsychology. *Statistical Science, 6,* 363–403.

Walker, E. H. (1974). Consciousness and Quantum Theory. In J. White (Ed.), *Psychic Exploration: A Challenge for Science* (pp. 544–568). New York: Putnam.

White, R. A. (1990). An Experience-centered Approach to Parapsychology. *Exceptional Human Experience, 8,* 7–36.

Wirth, D. P. (1990). The Effect of Non-contact Therapeutic Touch on the Healing Rate of Full Thickness Dermal Wounds. *Subtle Energies, 1,* 1–20.

Name Index

Subject Index